Solving Equations with Physical Understanding

Solving Equations with Physical Understanding

J R Acton and P T Squire

Adam Hilger Ltd, Bristol and Boston

British Library Cataloguing in Publication Data

Acton, J. R.
Solving equations with physical understanding.
1. Differential equations 2. Mathematical physics
I. Title II. Squire, P. T.
530.1′5535 QC20.7.D5

ISBN 0-85274-757-8 (hbk)
ISBN 0-85274-799-3 (pbk)

Published by Adam Hilger Ltd
Techno House, Redcliffe Way, Bristol BS1 6NX, England.
P O Box 230, Accord, MA 02018, USA.

Printed in Great Britain by J W Arrowsmith Ltd, Bristol

Contents

* Sections marked with an asterisk may be omitted at a first reading.

Preface

This book is written for three groups of readers:

(a) Students of physics, engineering and other applied sciences who have completed a conventional course in differential equations. Many of these students may realise that the subject is of great practical importance but have no confidence that they will be able to master the formal mathematics required.

(b) Professional scientists and engineers who were once in the former category, but now realise that mathematical models of real systems do throw up differential equations; and that such equations have a nasty habit of being different from the textbook examples. After an hour or so grappling with his old notes and a pile of authoritative but bewildering treatises, the unhappy member of this group goes in search of a mathematician or uses a computer and standard software that he only partly understands. In either case he has isolated himself from the process of solving the equation, and as a result has diminished his understanding of what the formal or computed solutions mean.

(c) Teachers who need to explain the physical significance of equations and their solutions in plain language to students of applied science and engineering whose interests and skills are more often in experimental work than in mathematics.

The common problem for people in these three groups is that formal mathematics is an obstacle rather than an aid to understanding.

While it has long been recognised that the ability to make order-of-magnitude numerical estimates is an essential skill of the working scientist and engineer, there is no corresponding method for a 'back-of-envelope' treatment of differential equations. What is needed is a method that bypasses the mathematical difficulties and also emphasises those features of the solution most useful in the design of practical devices and systems. It is the prime purpose of this book to fill this need.

The reader will be relieved to find that this does not require the use of any formal mathematics beyond that learned in sixth-form or first-year university courses. Instead, much greater use is made of physical and intuitive arguments of a kind familiar to experimental scientists. We shall often appeal to analogy, prior experience and experimental evidence, and make use of inference and hypothesis. Using this approach, the reader who cannot afford to carry a great weight of specialised mathematical equipment will find he can understand, and therefore enjoy, mathematical models whose equations are too difficult to solve by other means.

Chapter 1

Introduction

1.1 Approximation with understanding as its goal

Mathematical models usually describe either processes or devices. In processes, the system is changing with time; in devices, the changes studied may also be with time, or the interest may be in the way in which variables such as temperature or concentration change with position within the device. In both cases, rates of change are involved, so the mathematical model leads to a differential equation. The principal object of this book is to solve differential equations so as to obtain a broad understanding of the systems they describe, using a minimum of formal mathematics.

Differential equations may be classified in order of increasing difficulty:

(a) Very simple equations with simple and well understood solutions.

(b) Equations that can be solved exactly, but whose solutions are so complicated, or contain such unfamiliar functions, that only a specialist can understand their physical significance.

(c) Equations that cannot be solved at all by formal mathematics. There are no useful exact solution methods, for instance, for the majority of nonlinear equations.

The equations used for teaching and examination purposes are deliberately chosen mainly from the first class, with occasional examples from the second. Unfortunately, many equations arising from practical problems belong to the second or third classes, and their formal solution is either impossible or physically obscure.

The key to physical understanding of these equations is the use of approximation. Exact solutions are complicated and may need unfamiliar functions because, by definition, they must describe exactly every detail of the changes in the variables. By contrast, approximate solutions can strip away the overlying detail to show the essential relationships between the physical variables. What is more, these relationships can be expressed in symbols and words that are familiar to all scientists and engineers.

As an illustration, consider an equation that can be used as a mathematical model for the chemical reaction:

$$A + 2B \rightarrow C.$$

The equation is

$$\frac{dn}{dt} = K(a-n)(b-2n)^2 \tag{1.1}$$

where a and b are the numbers of molecules A and B at $t=0$, and n is the number of molecules C after time t.

The exact solution is

$$t = \frac{1}{K}\left[\frac{1}{2a-b}\left(\frac{1}{b-2n} - \frac{1}{b}\right) + \frac{2}{(2a-b)^2}\ln\left(\frac{1-2n/b}{1-n/a}\right)\right]. \tag{1.2}$$

This result is too complicated to give a direct answer to the most practical questions about the reaction, such as 'What does the curve of n against t look like?' and 'How does the shape of this curve depend upon a, b and K, the constants in the equation?' By contrast, the approximate solution† (assuming the reaction ends when all the molecules B are used up) answers these questions directly. It is

$$n \approx \tfrac{1}{2}b(1 - e^{-t/\tau}) \tag{1.3}$$

where

$$\tau = \frac{1}{Kab(1 - b/4a)}. \tag{1.4}$$

In words, solution (1.3) states that 'n increases approximately exponentially from zero at $t=0$ to a final value $n = \frac{1}{2}b$, with a time constant given by formula (1.4)'.

The exponential curve and its time constant are familiar ideas in science and engineering, and the curve of n against t (figure 1.1) can be pictured and sketched immediately, using either the mathematical description (1.3) or its verbal equivalent.

The answer to the second question, about the effect of changing a, b and K, is given in the compact formula (1.4) for the time constant τ. In words, 'τ is inversely proportional to a, b and K, subject to a correction that has a relatively slow variation with the ratio b/a'. Formulae like (1.4), which show the way in which the most important parameters of the system depend upon each other, will be called *design formulae*, since they embody the essential relationships needed by engineers who have to design practical systems.

The example shows what is meant by solving for physical understanding. What is required first is a solution that can be pictured graphically and described in simple and familiar words; the second requirement is a design formula showing the most important relationship between the parameters. In

† Derived in problem 3.6.

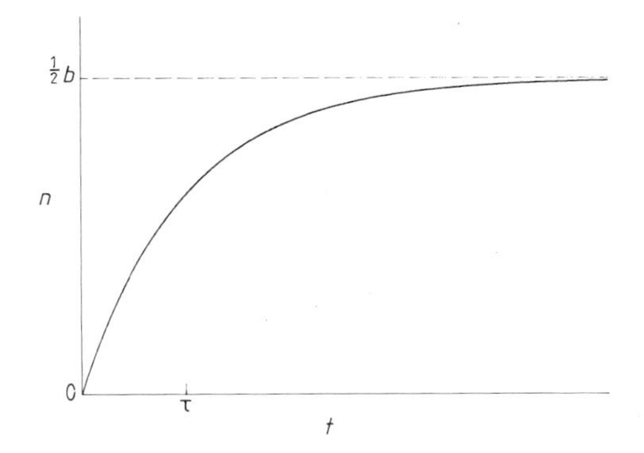

Figure 1.1 Approximate solution of equation (1.1).

the conventional methods of solution, a mathematical solution is obtained first, and from this graphs can be plotted and design formulae derived. This means that the graphs and the design formulae, which may be quite simple to understand, are obtained as secondary products of the primary mathematical investigation, which may be very complicated.

The special feature of the approximate method of solution used in this book is that it reverses this procedure. The first step is to use physical rather than mathematical arguments to make a qualitative sketch (QS for short) of the likely form of the graphical solution; the second step is to use the method of trial function (or TF) approximation to add more detail to the graph, and also to find the design formula directly. This qualitative sketch/trial function (QSTF) method thus goes straight to what is needed for physical understanding, instead of using the conventional roundabout route. The consequence is that the method is generally quicker and involves fewer mathematical steps than exact procedures. A further consequence is that, by avoiding the formal mathematical difficulties, this approximate approach works for many equations that are impossible to solve exactly in closed form.

1.2 Trial function approximation

Since trial function approximation, which is the second step of the QSTF method, may be unfamiliar, we shall first describe it in brief outline, leaving all the details of the steps to be filled in later. It is started by choosing a mathematical function, called the *trial function* (or TF for short), which is believed to be a good approximation to the exact solution; this TF always contains one or more *unknown parameters*. The second step is to combine the TF and the equation, using a process called *residual minimisation*, to find formulae for the unknown parameters in terms of the physical constants of the system being studied.

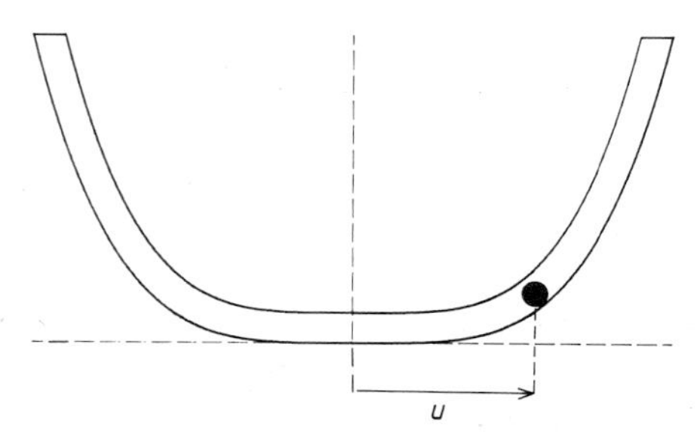

Figure 1.2 A ball-bearing oscillating in a smooth tube bent to produce a restoring force proportional to the cube of the displacement.

As an illustration, consider the motion of a ball-bearing oscillating in a glass tube that is bent into a curve such that the restoring force depends upon the cube of the displacement u (figure 1.2). The governing equation, ignoring frictional losses, is

$$\frac{d^2u}{dt^2}+cu^3=0 \tag{1.5}$$

and the auxiliary conditions are that the ball-bearing is released from rest at a displacement u_0 when $t=0$. Expressed mathematically, this is

$$u(0)=u_0 \qquad \underset{\text{[initial velocity zero]}}{u'(0)=0.} \tag{1.6}$$

The goal of solving this mathematical model is to find how the frequency f of the oscillation depends upon the physical parameters; in other words, the goal is a design formula for f. The methods for choosing trial functions will be explained later. In this case the TF used is

$$u^*=u_0 \cos \omega t. \tag{1.7}$$

This can be seen to fit the auxiliary conditions (1.6) and to be a physically reasonable way of describing an oscillation of amplitude u_0. The unknown parameter is the angular velocity ω.

The second stage is to find an approximation for this unknown parameter by *residual minimisation.* Again the details of this process will be discussed in great detail later; for the present, the important point is that it involves only a few lines of simple mathematics to reach the result

$$\omega=0.87c^{1/2}u_0. \tag{1.8}$$

The relationship between frequency and angular velocity is

$$f=\omega/2\pi \tag{1.9}$$

so the design formula for f is

$$f = \frac{0.87}{2\pi} c^{1/2} u_0 = 0.14 c^{1/2} u_0. \qquad (1.10)$$

This is the required design formula relating the most important characteristic of the performance of the oscillator, i.e. frequency f, to the physical constants c and u_0 present in the mathematical model.

The directness and simplicity of TF approximation contrasts with the tediousness of the formal method of solution. First, the exact solution of the equation, involving elliptic functions, would have to be found; then the rather unusual periodic properties of this solution recognised; and finally, not without some difficulty, the period found from appropriate tables. After all this, substantially the same design formula is obtained, differing only in that the numerical constant would be 0.135 instead of 0.14.

1.3 The qualitative sketch

The brief account of the TF approximation in the last section will have raised many questions. The first step, for instance, is to choose a TF that is believed to be a good approximation to the exact solution, and it is by no means clear how this is to be done. The rather scattered descriptions of the TF method in the literature are of little help; they do not describe any general technique for choosing a TF, and physical understanding is seldom their prime objective. We shall show in this and the next section that, for physical equations that describe real processes or devices, there is a systematic method for finding a suitable trial function.

The key to the problem is that for physical equations it is almost always possible to sketch a graph of the solution without using the equation at all. In the case of the chemical reaction discussed in §1.1, for instance, it is clear without any mathematics that the number n of molecules C must start from zero and increase with time. Intuition also suggests that the curve of n against t must be smooth, since there are no mechanisms to cause any humps or sudden changes in direction. Finally it is clear that the number of molecules C is eventually limited by exhaustion of one of the reactant molecules A or B. Putting these three statements together gives a picture of a smooth curve rising from $n=0$ and eventually flattening off at a steady value. When such a curve is sketched its shape must look like figure 1.1. Without any reference to the equation, therefore, we have been able to sketch the approximate shape of the solution curve, using only the initial conditions and physical commonsense.

A sketch of the expected solution made in this way without formally solving the equation will be called a *qualitative sketch*. It is drawn by first carefully considering any auxiliary conditions, which will usually fix the beginning and end of the curve, and then using physical intuition to fill in the intervening curve. In Chapter 2 and in later examples we shall show how this physical intuition can be supplemented by analogy with known solutions of simpler equations and direct inspection (without formal solution) of the equation itself.

The drawing of the qualitative sketch is the most important single step in solving with physical understanding. It serves three purposes:

(a) It goes a long way to answering the question. 'What does the solution look like?'

(b) It forces the solver to think in the first instance about the physics of the real apparatus, rather than about the abstract mathematical equation.

(c) As we shall show in the next section, it provides a systematic way of choosing the TF.

Two more ingredients are needed to complete our physical understanding, as defined in §1.1. First, although the qualitative sketch gives the shape of the curve, the scale of one or other of the axes is usually missing. Secondly, the appropriate design formula for the problem, relating the performance to the constants of the system, is wanted. Both these needs are supplied by the TF approximation, and we shall next show how the qualitative sketch makes the choice of TF more systematic. The two stages, the making of the qualitative sketch (or QS), followed by the TF approximation, together make up the QSTF method of solving with physical understanding.

1.4 Standard functions

When the qualitative sketches are made for a number of practical problems chosen from various fields, it turns out that they almost always look like one of the curves in figure 1.3(*a*)–(*e*). These curves may be classed as either sinusoidal (figure 1.3(*a*)), exponential (figure 1.3(*b*) and (*c*)) or parabolic (figure 1.3(*d*) and (*e*)). For some less common auxiliary conditions, the curves are displaced. An example is shown in figure 1.3(*f*), which is just the curve of figure 1.3(*b*) displaced upwards and to the right. The three shapes are so distinctive that they can be immediately recognised even when they are displaced, and the curve of figure 1.3(*f*) can be classed as exponential without any difficulty. There are good physical reasons why these graphical shapes are so common as solutions to physical equations. The sinusoid, with its repeated maxima and minima, is the simplest curve that can represent an oscillation of constant amplitude and frequency, and such oscillations occur in every branch of science. Similarly, the exponential, characterised by its long plateau, is the simplest curve that can describe a smooth change ending by becoming asymptotic to a steady state, behaviour typical of changes of state occurring in many different sciences. Finally, the parabola is the simplest curve characterised by symmetry about a single extremum, and such symmetry is a common factor in a whole variety of apparatus and devices. This classification of

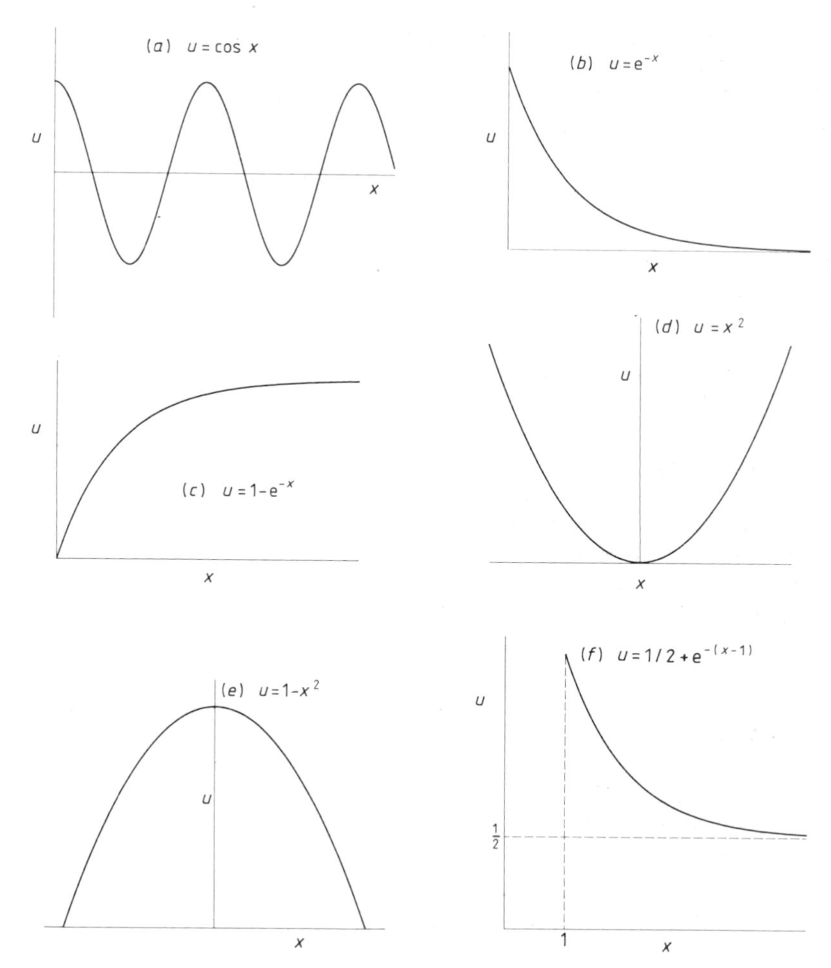

Figure 1.3 Curves of familiar functions that commonly occur as the solutions to physical equations: (*a*) sinusoidal, (*b*) and (*c*) exponential, (*d*) and (*e*) parabolic, and (*f*) displaced exponential.

qualitative sketches into one of three classes gives a systematic basis for the choice of TF.

The functions used to generate the curves in figure 1.3 cannot be used directly as TFs since they do not contain any unknown parameters, which are essential for the final step of the TF approximation. The simplest useful forms of TF corresponding to the three kinds of curve are:

sinusoidal

$$u = A \cos \omega t \tag{1.11}$$

exponential

$$u = A\,\mathrm{e}^{-t/\tau} \qquad u = A(1 - \mathrm{e}^{-t/\tau}) \tag{1.12}$$

parabolic

$$u = A(1 - x^2/l^2). \tag{1.13}$$

These each contain two parameters that have a clear physical meaning: A is the amplitude of the cosine wave and for the exponential and the parabola it is the height of the curve; ω, τ and l respectively correspond to the frequency ($f = \omega/2\pi$) of an oscillation, the time constant of a change and the half-width of a symmetrical curve. If the qualitative sketch is displaced, then constants must be added to these simple trial functions to shift the origin of the axes. The displaced curve of figure 1.3(f), for instance, corresponds to

$$u = u_\infty + A\,\mathrm{e}^{-(t - t_0)/\tau}$$

where $u_\infty = \frac{1}{2}$ and $t_0 = 1$ are constants describing the shift of origin. When examples are studied, it will be seen that such additional constants always have a clear physical interpretation.

The cosine, exponential and parabola, in either their simple or displaced forms, will be called the three standard functions of approximation or more simply the *standard functions*. They form the basis both for studying the QSTF method and for using it to solve equations. The technique of the QSTF method is best learnt by first mastering problems for which the standard functions can be used as TFs. This is the method followed in this book, in which the technique is first given with worked examples for exponential TFs (Chapters 3 and 4), and then for parabolic (Chapter 5) and cosine TFs (Chapters 6 and 7). Once the techniques for the standard functions are mastered for ordinary differential equations, it is possible to extend their use to partial differential equations (Chapters 8–10).

At the end of each chapter a number of exercises (for practising techniques) and problems (for applying them) are given. Like the examples in the main text, these are carefully chosen to avoid redundancy, and they often introduce new points or extensions of the method. Some of the working in the later chapters presupposes that examples in earlier chapters have been worked through; in particular, it is impossible to understand the chapters on partial differential equations without a firm grasp of the QSTF method applied to ordinary differential equations.

In the examples, the qualitative sketch is first drawn and recognised as having one of the three standard shapes; it is then compared with graphs of the corresponding standard functions in their simple and displaced forms. The graph that matches the qualitative sketch is picked out and the corresponding function is the TF for the problem. This procedure automatically leads to useful results. The qualitative sketch ensures that the corresponding TF is physically reasonable and conforms to the auxiliary conditions. The use of the standard functions means that any mathematics will be relatively straightforward and familiar, and therefore clearly understood. Finally, because the standard functions contain only the most important parameters, their use guarantees that the TF approximation will yield a design formula of direct practical importance.

As mentioned in §1.2, the last stage of the TF approximation is residual minimisation. This stage can be made quite simple, but the technique is best learnt by following some examples, and is therefore first introduced in Chapter 3. The techniques for the three standard functions differ only in small details, so that the method becomes more and more familiar as more equations are studied. This is in direct contrast to the methods of formal mathematics, in which new and often highly specialised techniques and functions have to be learnt as each new type of physical equation is encountered.

Once the QSTF method using standard functions is mastered, it is possible to tackle some of the less common problems that need nonstandard TFs (see Chapter 11). Very often, however, the standard TFs can be adapted slightly to meet the new circumstances, and a number of examples of such adaptations are given in the appropriate chapters alongside standard problems.

1.5 Useful accuracy

It is seldom in applied science or engineering that exact solutions are required. This is for two reasons. The first is that the equation itself is not an exact description of the system that it models. Some terms may be omitted because they are 'small', which may mean anything from 10^{-6} to one-third of the total. Other terms are approximate because they include physical parameters or functions that are themselves approximate. For example, in an equation of heat power balance, loss by radiation and conduction might be neglected, while the convective loss is described by a term based on an empirical law of dubious validity, involving a parameter—the heat transfer coefficient—which is not known accurately and which depends on several uncontrolled factors.

The second reason why exact solutions are seldom required by the applied scientist or engineer is that the use to which the solutions are to be put does not warrant it. One of the arts of good applied science is to achieve accuracy *adequate for a particular purpose.* The pursuit of excessive accuracy is both time-consuming and costly. It is a strange fact that this is recognised in experimental science, where one more significant figure may increase the cost of a measurement 10-fold, whereas, in calculations performed on the same system, much time is often wasted in seeking accurate solutions to inaccurate governing equations. It is as if the mathematical form in which the problem is expressed exerted some kind of hypnotic influence over the solver, tempting him to pursue solutions that are mathematically respectable but physically unrealistic or even meaningless.

In the early stages of experimental work, particularly in the course of technological invention, research and development, therefore, it is unreasonable and uneconomic to insist on even 1% accuracy. On the other hand, an error by a factor of 2 or 3 could be misleading. These lower and upper extremes suggest that errors up to about 30% at worst would be acceptable; the typical accuracy would then be around 10%.

In the next stage of device evolution, early development, the design formulae are used to adjust the performance as experimental knowledge increases. The requirement here is not absolute numerical accuracy but a good description of the way in which changes in the physical constants of the device can be used to change performance. For the chemical reaction considered in §1.1, for instance, the knowledge that the time constant (formula (1.4)) is inversely proportional to K, a and b would be more useful in developing the process than the precise value of the numerical constant. Once again, provided the functional dependence is well described, a numerical accuracy of about 10% is adequate.

It is only in the final stages of development, or in the routine design of devices that are well understood, that precise numerical values may be needed to optimise the design or to fill in details which are not made clear by approximate treatment. This high numerical precision is usually obtained using a computer, and here a great deal of time and effort can be saved if the computation is preceded by an approximate solution.

In this book an approximation will be called *useful* if the error is expected to lie within the limits $\pm 30\%$, which means a typical error of around 10%. The TF approximation based on standard and closely related functions chosen by the physical considerations outlined in §§1.3 and 1.4 normally produces design formulae having numerical accuracy well within these limits. The reader will find this substantiated when he studies the examples.

1.6 Scope and aims of the book

A book on solving equations with physical understanding can only cover a proportion of the equation types that may arise as mathematical models of physical devices or systems. Apart from the limitations of space, there are for instance equations that involve so many variables that they can only be treated by using condensed notations such as matrix or tensor algebra. It is doubtful whether such equations can be understood in the direct physical way that is the object of this book, except possibly by specialists who spend a long time becoming familiar with the mathematics necessary in a particular subject area.

The problems studied here are therefore mostly limited to those modelled by single ordinary differential equations (Chapters 2–7) and to relatively simple partial differential equations with only two independent variables (Chapters 8–10).

The main aims of the book are:

(a) To illustrate the solutions, mainly in the form of design formulae, of about 50 problems selected for the most part from practical experience. These problems have been chosen so that the equations are almost all different, and between them they cover many of the most common types, very few of which are discussed in standard introductory treatments.

(b) To provide, in the course of working the examples in the text, a fairly complete survey of the systematic qualitative sketch/trial function or QSTF method outlined in the previous sections; and where necessary to point out possible difficulties and limitations. The concepts and techniques are studied to a depth sufficient to allow the reader to extend the method to equations beyond those discussed in the text. It is our hope that the reader will therefore be encouraged to be adventurous in the application of the methods described to his own field of study.

(c) To use the QSTF method to obtain physical insight into a number of mathematical problems that are common sources of difficulty to non-specialists: nonlinear equations (about which introductory texts are almost silent) and in particular nonlinear oscillations; eigenvalue problems, whose formal treatment is often so complicated that their physical significance is lost; the commonest singularities, which are either not mentioned at all in elementary texts or discussed so formally that their physical significance is obscured; and conduction and diffusion problems, whose conventional solutions in terms of oscillatory eigenfunctions are often unnecessarily complicated and physically incongruous.

The mathematical texts that do treat these subjects are often advanced, and make difficult reading for the applied scientist and engineer. These texts are also by their nature more concerned with rigour and completeness than with physical understanding. Our object is not to replace such texts but to provide a preliminary understanding of the most important physical aspects of the equations and their solutions.

Chapter 2

Mathematical preliminaries

2.1 Conventions and abbreviations

In this chapter we shall review briefly the formal solutions of the very simplest ordinary differential equations, at a depth just sufficient to cover the needs of later chapters. We shall also describe how to identify oscillatory equations, review the concept of first-order approximations and introduce certain helpful conventions that will be used throughout the book but are not in general use. Finally, the problems at the end of the chapter provide a complete revision course in the small amount of basic mathematics needed for the QSTF method.

Some conventions and abbreviations must be mentioned immediately. First, to emphasise that an equation describes a balance between physical effects, we shall often write a verbal statement underneath the governing equation. For example, Newton's second law of motion expresses the balance between applied force and mass times acceleration for a particle. In symbols and words, it may be written

$$m\frac{\mathrm{d}v}{\mathrm{d}t} \quad = \quad F.$$

[mass × acceleration = applied force]

The second convention is that, in any equation or formula quoted in this book, it is assumed that all the quantities are real, and that all constants and parameters are positive, unless otherwise stated. This convention would be useless in studying abstract equations formally, but is very useful for physical equations, since all experimentally observable quantities are represented by real numbers, and physical properties such as mass and resistance are almost always positive quantities. The third convention, concerning 'small' quantities, is explained in §2.8.

Besides the abbreviation TF for trial function, we shall use the following to denote the various classes of differential equation:

ODE ordinary differential equation,
ODE 1 first-order ordinary differential equation,
ODE 2 second-order ordinary differential equation,
PDE partial differential equation.

We shall also use LHS and RHS to denote the left-hand and right-hand sides of an equation, and primes to denote ordinary differentiation, such as

$$u' = \frac{du}{dx} \qquad u'' = \frac{d^2u}{dx^2}.$$

2.2 Definition of terms used to describe ordinary differential equations

An *ordinary differential equation* is an equation containing one or more derivatives of a *dependent variable* with respect to a single *independent variable.* For example,

$$\frac{du}{dx} + u = 1 \tag{2.1}$$

contains the first derivative of the dependent variable u with respect to the independent variable x.

A formal *explicit solution* of an ordinary differential equation is an explicit relation between the dependent and independent variables, that is

$$u = f(x)$$

or its inverse,

$$x = g(u).$$

The *order* of an ordinary differential equation is the order of the highest derivative present. Thus equation (2.1) is of the first order, whereas the equation

$$\frac{d^2u}{dx^2} + u = \sin \omega x \tag{2.2}$$

is of the second order.

An ordinary differential equation is said to be *linear* if the dependent variable occurs no more than once in each term of the equation and neither it nor any of its derivatives is raised to a power other than unity. Otherwise it is *nonlinear.* For instance, both equations (2.1) and (2.2) are linear, but the following equations are nonlinear:

$$\frac{du}{dx} + u^{1.6} = 1 \qquad \text{(power of } u \text{ other than 1)} \tag{2.3}$$

and

$$u\frac{du}{dx} + u = 1 \qquad (u \text{ occurs twice in first term}). \tag{2.4}$$

The quantity on the RHS of an ordinary differential equation is called the *forcing term*; it is denoted in this book by $F(x)$, where x is the independent variable. The forcing term must not contain the dependent variable u; it may be a positive or negative constant or any function of x such as $\sin \omega x$.

The factors multiplying the independent variable or its derivatives in each term are called the *coefficients*. In all the equations (2.1)–(2.4) the coefficients are unity. If the coefficients in an equation are all constants, the equation is classified as being of the *constant coefficient* type. If any coefficient contains the independent variable, it is of the *nonconstant coefficient* or *variable coefficient* type. An example of this latter type is the equation

$$\frac{\mathrm{d}u}{\mathrm{d}x}+(1+b\sin\omega x)u=1 \tag{2.5}$$

in which the coefficient $(1+b\sin\omega x)$ is not constant. Notice that in this equation b is an example of a *parameter*, which by our convention must be positive; the variables u and x are not subject to this convention and can be either positive or negative.

A first-order ordinary differential equation is said to be *separable* if it can be written in the form

$$F(u)\,\mathrm{d}u=G(x)\,\mathrm{d}x. \tag{2.6}$$

For example, equation (2.1) is separable, as it may be rearranged into the form

$$\frac{\mathrm{d}u}{1-u}=\mathrm{d}x.$$

The reason for trying to write a differential equation in this form is that it can then be solved, at least in principle, by direct integration. The solution is expressed formally by the relation

$$\int F(u)\,\mathrm{d}u=\int G(x)\,\mathrm{d}x+\text{constant} \tag{2.7}$$

which no longer contains derivatives and is in that sense a simpler equation. However, the integrations may not be possible in closed form, so that an explicit solution in the form $u=f(x)$ may not be possible.

2.3 Exact solutions of simple linear first-order ordinary differential equations

There are two common and very simple linear ODE 1 whose exact solutions should be familiar. The first is

$$\frac{\mathrm{d}u}{\mathrm{d}t}+au=0 \tag{2.8}$$

whose solution, for the initial condition that $u=u_0$ at $t=0$, is

$$u=u_0\,\mathrm{e}^{-t/\tau}. \tag{2.9}$$

Here τ is called the *time constant*, and is given by

$$\tau=1/a.$$

The second equation is similar, but has a constant forcing term c, that is

$$\frac{\mathrm{d}u}{\mathrm{d}t}+au=c. \tag{2.10}$$

Its solution, for the initial condition $u=0$ at $t=0$, is

$$u=\frac{c}{a}(1-\mathrm{e}^{-t/\tau}) \tag{2.11}$$

with the time constant τ again equal to $1/a$.

The solutions (2.9) and (2.11) are shown graphically in figure 2.1(a) and (b). In words, solution (2.9) describes an exponential fall from $u=u_0$ at $t=0$ to $u=0$ when $t\to\infty$, and solution (2.11) describes an exponential rise from $u=0$ to a final value of $u_\infty=c/a$.

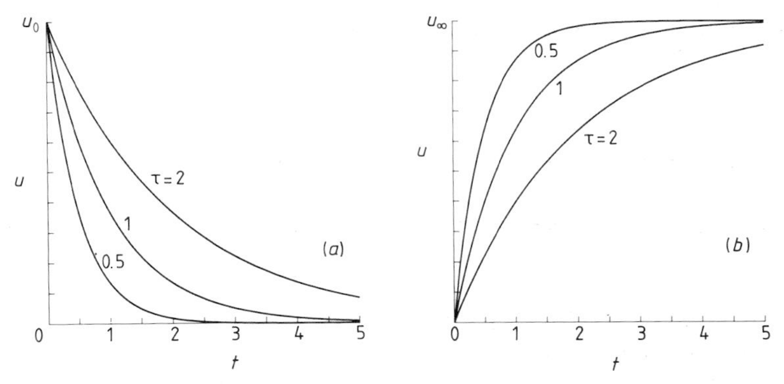

Figure 2.1 Graphs of (a) solution (2.9) of equation (2.8) and (b) solution (2.11) of equation (2.10) (curves correspond to $\tau=1/a=0.5$, 1 and 2).

From the point of view of physical understanding, it is important to recognise at sight that linear ODE 1 like equations (2.8) and (2.10), with constant coefficient a and constant or zero forcing terms, describe exponential changes whose time constant can be written down immediately as

$$\tau=1/a. \tag{2.12}$$

The time constant is a familiar way of fixing the timescale of the change, and is defined as the time to complete a fraction (e − 1)/e (or about 0.63) of the change in u.

The relation (2.12) is a very simple example of a design formula, defined in §1.2, since it is a relation between a parameter describing the behaviour of the system (e.g. τ) and the constant present in the mathematical model (e.g. the constant a in the equation).

Another point about these equations, which will be applied more generally in later chapters, is that when u reaches its final steady state the curve has ended in a flat plateau, and $\mathrm{d}u/\mathrm{d}t$ must be zero. The final value of u when $t \to \infty$ can therefore always be found by putting $\mathrm{d}u/\mathrm{d}t = 0$ in the equation. For equation (2.10), for instance, this gives

$$au_{\infty} = c$$

whence

$$u_{\infty} = c/a.$$

2.4 Exact solutions of simple linear second-order ordinary differential equations

Two common ODE 2 whose exact solutions are important for physical understanding both have two linear terms with constant coefficients. They are

$$\frac{\mathrm{d}^2 u}{\mathrm{d}t^2} - k^2 u = 0 \qquad (2.13)$$

and

$$\frac{\mathrm{d}^2 u}{\mathrm{d}t^2} + \omega_0^2 u = 0. \qquad (2.14)$$

The crucial difference between these two very similar equations is the sign of the coefficient of u in the second term. This determines whether the solutions are exponential or oscillatory. In equation (2.13) the coefficient is negative, and the general solution is

$$u = A\,\mathrm{e}^{-kt} + B\,\mathrm{e}^{+kt}. \qquad (2.15)$$

This solution is an example of the very important *principle of superposition*, which states that for a *linear* equation any linear combination of separate solutions is also a solution. In this case e^{-kt} and e^{+kt} are both solutions of equation (2.13), as may be verified by substitution. The solution (2.15) is a linear combination of these solutions, in which A and B are constants that must be determined from the *auxiliary conditions*, such as initial or boundary conditions.

For most problems in time, the general solution (2.15) can be reduced to a single term by the following argument. For physical equations modelling real systems, u is a physically observable quantity, such as temperature or concentration, which must remain finite. Since the positive exponential in the second term of solution (2.15) becomes infinite when $t \to \infty$, B must be zero to prevent u becoming infinite; the solution is therefore simply

$$u = A\,\mathrm{e}^{-t/\tau} \tag{2.16}$$

where

$$\tau = 1/k. \tag{2.17}$$

This describes an exponential fall from $u=A$ at $t=0$ to $u=0$ as $t \to \infty$, with a time constant equal to $1/k$.

Second-order equations of the form (2.13) also occur in which the variation of u depends upon changes in distance x rather than time t. The auxiliary conditions for these problems are discussed in Chapters 4 and 5.

The second equation (2.14) has a positive coefficient of u, and in this case the general solution is

$$u = C\cos\omega_0 t + D\sin\omega_0 t. \tag{2.18a}$$

This solution describes an oscillation at the *angular velocity* ω_0. Once again C and D are constants to be determined from the auxiliary conditions. Physically the solution is clearer if it is written in the mathematically equivalent form (appendix 6)

$$u = A\cos(\omega_0 t + \phi) \tag{2.18b}$$

where A and ϕ are now the constants to be determined from the auxiliary conditions. This still describes an oscillation at angular velocity ω_0, but now the constants have a direct physical meaning, A being the *amplitude* of the oscillations and ϕ being the *phase*.

The angular velocity ω_0 is related to the *frequency* f and *period* T of the oscillation by the formulae

$$f = \omega_0/2\pi \tag{2.19}$$

and

$$T = 1/f = 2\pi/\omega_0. \tag{2.20}$$

These are further examples of simple design formulae relating the important characteristics of an oscillator (f and T) to the constant ω_0 in the physical equation. In this example the formulae are exact, since the solutions from which they are derived are exact solutions of the original equation.

For physical understanding, the most important feature of these equations is the distinction between the nonoscillatory equation (2.13) and the oscillatory equation (2.14), according to the negative or positive sign of the coefficient of u. After this, the most useful results are the simple design relations for τ, f and T in terms of k and ω_0.

The equations described in this section and in §2.3 are among the very few equations whose design formulae for τ, f and T are so simple that they can be written down immediately on inspection of the equation. Later chapters will contain frequent references to these simple equations, and familiarity with them and their solutions will be of the greatest help in understanding the QSTF solutions of more complicated examples.

2.5 Damped linear oscillations

An important modification of the simple harmonic oscillator equation (2.14) is the equation

$$\frac{d^2u}{dt^2}+p\frac{du}{dt}+\omega_0^2u=0. \tag{2.21}$$

This is still linear with constant coefficients, but differs from equation (2.14) by the presence of the *damping term* $p\,du/dt$, so called because it causes the amplitude of oscillation to decay. The physical origin of damping is exemplified by friction in a mechanical oscillator such as a pendulum, or resistance in an electrical oscillator. Many practical devices can be modelled approximately by this equation, whose general solution is

$$u=A\,e^{-pt/2}\cos(\omega_n t+\phi) \tag{2.22}$$

where A and ϕ are again the two constants to be chosen to satisfy the auxiliary conditions.

The solution (2.22) describes an oscillation of angular velocity ω_n and phase ϕ, but differs from the solution (2.18b) for the undamped (or free) oscillator in two important respects. The most obvious difference is that for damped oscillations the amplitude is no longer constant, but decays exponentially. Figure 2.2 shows graphs of the solution (2.22) for the same initial conditions and various values of the damping constant p. Figure 2.2(a) shows undamped oscillations ($p=0$), corresponding to equation (2.14). Figure 2.2(b) and (c) show the effect of progressively greater damping. Finally, figure 2.2(d) shows the solution of the *critically damped oscillator*, where the value of p is just large enough to prevent oscillations altogether.

The second difference is that the angular velocity ω_n of the oscillation is no longer equal to ω_0, but is reduced by damping. The design formula for ω_n is

$$\omega_n=\omega_0(1-\tfrac{1}{4}p^2/\omega_0^2)^{1/2} \tag{2.23}$$

from which the design formulae $f_n=\omega_n/2\pi$ and $T=2\pi/\omega_n$ can be derived. The increase in the period is too small to be obvious in the graphs of figure 2.2.

Formula (2.23) can be used to predict the transition from oscillatory to nonoscillatory behaviour. As p increases, ω_n decreases, and when

$$\omega_0^2-\tfrac{1}{4}p^2=0$$

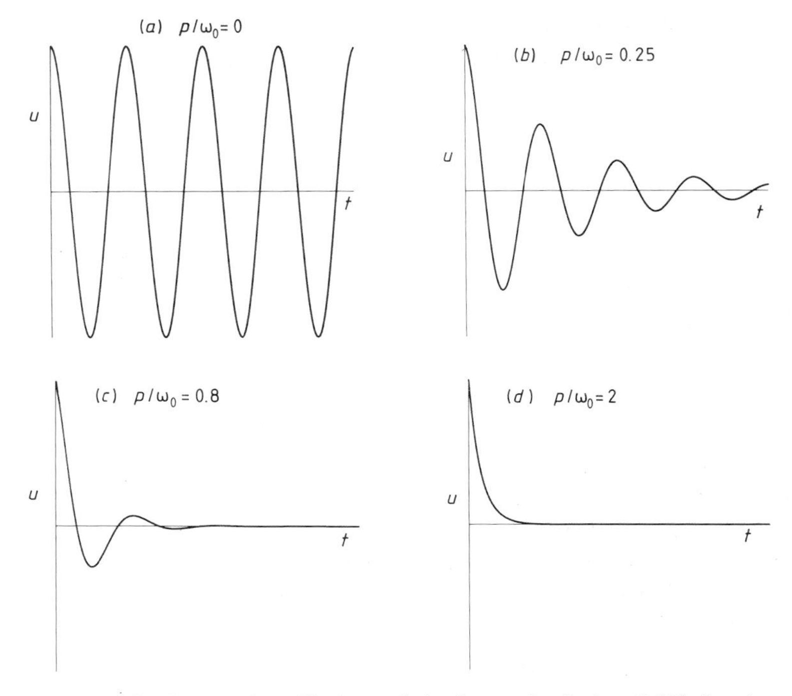

Figure 2.2 Damped oscillations of the form of solution (2.22), for the initial conditions $u(0)=1, u'(0)=0, \omega_0=1$. ω_n is given by formula (2.23): (*a*) undamped; (*b*) and (*c*) progressively greater damping; (*d*) critical damping.

ω_n becomes zero. This condition is called *critical damping*. Any further increase in p results in an imaginary frequency; that is, the system does not oscillate, and the solution is a decay curve like figure 2.2(*d*).

In the previous section, the criterion given for oscillatory behaviour was the sign of the coefficient of u. According to this rule, the positive sign of $+\omega_0^2$ in equation (2.21) indicates the possibility of oscillatory solutions, and we therefore describe this equation as *inherently oscillatory*. Whether or not oscillations actually occur is determined by the sign of the *discriminant*

$$\omega_0^2 - \tfrac{1}{4}p^2. \tag{2.24}$$

If the discriminant is positive, the solutions are oscillatory; otherwise they are nonoscillatory.

2.6 Free oscillations

As explained in Chapter 1, the TF method makes extensive use of the exponential and sinusoidal functions, which are two of the set of three standard functions. The decision as to which, if any, of these functions should be used can often be taken on physical grounds, but this is not always possible. In any case, it helps understanding if an equation can be classified as oscillatory by direct inspection of its terms.

For the simplest ODE 2 discussed in §2.4, the decision can be made immediately according to the sign of the coefficient of u; a positive sign signifies an oscillation. We shall now use physical arguments to show why essentially the same test can be applied to a more complicated class of equations, of the general form

$$\frac{d^2u}{dt^2}+qu=0. \tag{2.25}$$

The coefficient q is no longer necessarily a constant, but may be a function of t or u.

Equation (2.25) contains neither damping nor forcing terms; if it has an oscillatory solution, it is therefore a mathematical model for *free oscillators*, so called because they are free from the influences of either damping or forcing. The following argument explains how equations of this form can be identified as oscillatory.

A mechanical system will oscillate if a *restoring force* acts so as to accelerate the displaced object back towards the equilibrium position. The acceleration must therefore have the opposite sign to the displacement for most of the time. The best known example is the simple pendulum (figure 2.3), with the equation

$$\frac{d^2u}{dt^2} = -\left(\frac{g}{l}\right)u.$$

[acceleration $\propto$ − restoring force]

Since g/l is positive, the acceleration d^2u/dt^2 is always of opposite sign to the displacement u, causing the pendulum to swing back towards the centre.

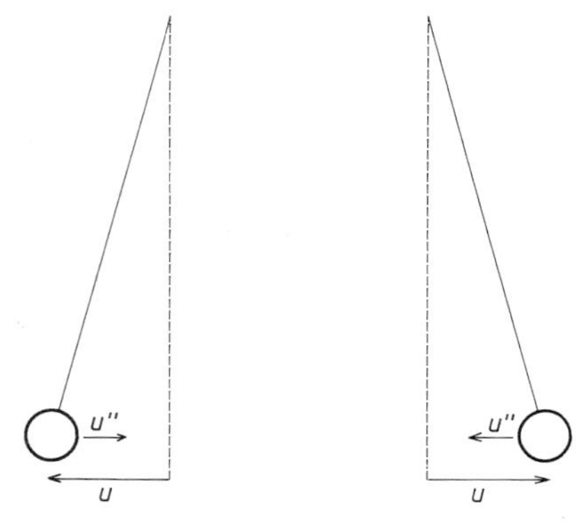

Figure 2.3 Simple pendulum, showing the opposite signs of the displacement u and acceleration u'' throughout the motion.

Similarly the nonlinear equation

$$\frac{d^2u}{dt^2}+c^2u^3=0$$

describes an oscillation: when the displacement u is positive, the acceleration

$$\frac{d^2u}{dt^2}=-c^2u^3 \qquad (2.26)$$

is negative; and conversely, when u is negative, the acceleration is positive.

By contrast the equation

$$\frac{d^2u}{dt^2}-c^2u^3=0$$

cannot describe an oscillation, for

$$u''=+c^2u^3$$

so the acceleration u'' has the same sign as the displacement u. It follows that the physical system cannot return to the central equilibrium position $u=0$, since the thing that is displaced is being accelerated even further away.

Quite generally, any ODE 2 of the form (2.25) will describe oscillations provided the coefficient q is positive. Coefficient q need not be a constant; provided it is positive throughout the motion, u'' will always have the opposite sign to u and the equation must describe an oscillation.

For example, the equation

$$\frac{d^2u}{dt^2}+kt^2u^5=0$$

may be written in the form

$$u''+(kt^2u^4)u=0$$

and the coefficient of u is seen to be

$$q=kt^2u^4.$$

Since k is positive (by the convention explained in §2.1) and t^2 and u^4 must be positive, q is positive and the equation therefore describes an oscillation.

Other forms of physical equations that describe free oscillations may be encountered, but the type of equation (2.25) is much the most common, and the simple sign test will be used in many examples discussed in later chapters.

2.7 Inherent oscillations

The possibility of natural oscillations can be recognised by inspection of equations even if they contain more than the two terms of the free oscillator equation (2.25). We shall define an *inherently oscillatory system* as any system that would oscillate freely in the absence of damping or forcing.

A method that will identify most inherently oscillatory equations is first to remove any damping and oscillatory forcing terms present†. If what remains is a free oscillator equation, then the original equation is inherently oscillatory.

For example, the equation

$$\frac{d^2u}{dt^2} + \underbrace{p\frac{du}{dt}}_{\text{damping term}} + u^3 = \underbrace{\sin \omega t}_{\text{oscillatory forcing term}} \tag{2.27}$$

when reduced to its undamped, unforced form, is

$$\frac{d^2u}{dt^2} + u^3 = 0$$

or in the form of equation (2.25),

$$u'' + (u^2)u = 0.$$

Since $q = u^2$, which is always positive, this is a free oscillator equation, and so the original equation (2.27) is inherently oscillatory.

Whether an inherent oscillator will actually oscillate depends on the effect of the neglected terms, and sometimes on the starting conditions. The importance of the concept is that it allows the possibility of oscillations to be detected. Once recognised, this possibility can then be investigated further using oscillatory TFS. Inherent oscillations in time are studied using sinusoidal TFS in Chapter 6 (unforced) and Chapter 7 (forced). Inherent oscillations in space usually correspond to eigenvalue problems, which commonly arise in the process of solving PDES (see Chapters 8–10).

2.8 Approximations based on smallness

Many otherwise difficult problems can be solved, and many formulae greatly simplified, if it is known that a parameter or variable is relatively small.

If a number x is small compared to unity, then x^2 is much smaller, and higher powers are smaller still. For instance, if $x = 10^{-2}$, then $x^2 = 10^{-4}$, $x^3 = 10^{-6}$, and so on. Consider next a *polynomial function* composed of a number of terms having successively higher powers of x, such as

$$y = 1 + x + 10x^2 + 10^3x^3.$$

† For equations with a constant forcing term, see §6.2.

When $x=0.1$ this becomes

$$y=1+0.1+0.1+1$$

and the last term is very important because of its relatively large coefficient. But if $x=10^{-3}$ the terms now read

$$y=1+10^{-3}+10^{-5}+10^{-6}$$

and the terms fall off successively in magnitude.

Quite generally, for any polynomial $P(x)$ or for any convergent power series, it is always possible to find a value of x that is sufficiently small to make the terms decrease in this way. It is then possible to approximate the polynomial by ignoring the later terms. Binomial functions will frequently be approximated in this way in the examples. For instance, if

$$y=\frac{1}{1+x}=(1+x)^{-1}$$

it may be expanded by the binomial theorem (appendix 2) as the series

$$y=1-x+x^2-x^3+\cdots$$

where $+\cdots$ indicates terms in yet higher powers of x.

If x is small enough, the contribution of each successive term diminishes, and the later terms can be neglected. The crudest approximation is to neglect all but the first term, and write

$$y=1.$$

This is called the *zero-order approximation* since all terms having powers of x higher than zero have been neglected. It is the value of the function obtained by putting $x=0$.

A better approximation is obtained by retaining two terms

$$y=1-x.$$

This is called the *approximation to first order in* x, since all terms having powers of x higher than x^1 have been neglected. Similarly the *approximation to second order in* x is

$$y=1-x^1+x^2$$

and so on.

Functionally the zero-order approximation is too crude, since it implies that y does not depend on x at all. From the point of view of physical understanding of the way y depends on x when x is small, we need the simplest approximation that contains x. For the binomial example, this is the first-order approximation

$$y=1-x.$$

For the function

$$y=\cos x=1-\frac{x^2}{2}+\frac{x^4}{24}+\cdots$$

the simplest approximation that contains x is the second-order approximation

$$y=1-\frac{x^2}{2}.$$

We shall call an approximation to a power series that retains only terms up to that with the lowest non-zero power of x an x-*small* approximation. For example, we shall frequently write

$$\frac{1}{1+x}=1-x \qquad (x \text{ small}).$$

By this we mean that $(1+x)^{-1}$ has been expanded by the binomial theorem and the unwanted higher terms omitted, on the basis that x is small enough to ensure that such an approximation will give numerically useful results.

In expressions containing several parameters, the approximation will depend on which parameter is small. Thus if

$$y=1+x+\beta x^2+\beta^2 x^3$$

then

$$y=1+x \qquad (x \text{ small})$$

but

$$y=(1+x)+\beta x^2 \qquad (\beta \text{ small}).$$

In the first case, terms up to the lowest power of x are retained; in the second case, terms up to the lowest power of β.

A working rule is that small approximations with two terms are useful provided the magnitude of the second term does not exceed about one-third of the first term. For example, if

$$y=\mathrm{e}^{-2x}$$

then

$$y=1-2x \qquad (x \text{ small}).$$

This should be useful up to $2x=\frac{1}{3}$, or $x=\frac{1}{6}$; the approximation then gives $y=0.67$, whereas the exact value is 0.72.

The simple rule just given cannot be expected to apply to every conceivable case. The most likely cause of breakdown in physical examples is if the terms of a series do not have alternating signs; some care should be exercised in approximating such series by the first two terms. However, alternating series are far more commonly found as solutions of physical equations, and the rule is a useful guide as to what is meant by 'x is small'.

Problems

(These problems are mostly revision or exercises to give practice in the techniques described in the chapter. The answers to most of them can be written down directly. References to appropriate parts of the text are given in square brackets where this is thought to be helpful.)

2.1

Classify the following equations according to their order (1 or 2) and as linear (L) or nonlinear (NL). [§2.2]

(a) $m\dfrac{du}{dt}=-bu+mg$

(b) $\dfrac{d^2u}{dx^2}+u^4=0$

(c) $\dfrac{d^2u}{dr^2}+\dfrac{1}{r}\dfrac{du}{dr}+bu=0 \qquad (r>0)$

(d) $\dfrac{d^2\theta}{dt^2}-2a\theta=0$

(e) $\dfrac{d^2u}{dx^2}+\dfrac{1}{u}=0$

(f) $\dfrac{dn}{dt}=-Kn$

(g) $\left(m_0+m_f-\dfrac{m_f t}{t_b}\right)\dfrac{dv}{dt}+Rv^2=T$

(h) $L\dfrac{d^2Q}{dt^2}+\dfrac{Q}{C}=0$

(i) $\dfrac{d^2u}{dt^2}+(1-au^2)\dfrac{du}{dt}+bu=0.$

2.2

(a) Rewrite equations (a) and (f) of problem 2.1 in standard form (highest differential, free of constants, must be the first term on LHS; forcing terms must be on RHS).

(b) Write down the design formula for the time constant τ of the changes modelled by equation (a) of problem 2.1. [Equation (2.10) and result (2.12)]

(c) Write down the design formula for the final steady-state values of u for equation (a) and n for equation (f) of problem 2.1. [End of §2.3]

2.3

(a) Decide whether equation (d) of problem 2.1 has an exponential or oscillatory solution, and write down a design formula for the time constant τ or frequency f, whichever is appropriate. [§2.4]

(b) Repeat for equation (h) of problem 2.1.

2.4

(a) Which of equations (b), (c), (e), (g) and (i) of problem 2.1 model a free oscillator? [§2.6]

(b) Which of these equations model systems that are inherently oscillatory but not free oscillators?

2.5

Show that, according to the convention of §2.8:

(a) $\sin x = x$ (x small)

(b) $e^{-kx} = 1 - kx$ (x small)

(c) $\dfrac{\sin \omega x}{x} = \omega - \tfrac{1}{6}\omega^3 x^2$ (x small)

(d) $\dfrac{\sin \omega x}{x} = \omega$ (ω small)

(e) $(1+k)(1+4k) = 1 + 5k$ (k small)

[Use expansions in appendix 2 as necessary]

(f) $\dfrac{1+b}{1-b} = 1 + 2b$ (b small)

[Expand $(1-b)^{-1}$]

(g) $(1+x)\,e^{-x} = 1 - \dfrac{x^2}{2}$ (x small)

2.6

(a) Show that the largest value of k for which you would expect the approximation in problem 2.5(e) to be useful is $k = \frac{1}{15}$; the approximation then gives the value 1.33. Use a calculator to show that the exact value in this case is 1.35. [End of §2.8]

(b) For the function in problem 2.5(g) show that the largest value of x for which the approximation should be valid is $x^2 = \frac{2}{3}$; show that for this value of x

the approximate and exact (by calculator) values are 0.67 and 0.80 respectively.

(These illustrate relatively favourable and unfavourable cases, the errors being less than 2% for (a) and nearly 20% for (b). Errors of this kind, averaging about 10%, and very rarely exceeding 30%, are typical of the application of the rule given in §2.8.)

2.7

If

$$y = 1 + a\omega^2 \cos \omega x + a^2 \sin \omega x$$

write down the small approximations to y for the following conditions:

(a) x small;
(b) ω small;
(c) a small.

[Appendix 2]

2.8

If a sphere of radius R, floating half-immersed in a liquid, is disturbed, it oscillates up and down. The equation for the change in depth h, measured from the halfway position, is

$$\frac{d^2h}{dt^2} + \frac{3g}{2R}\left(h - \frac{h^3}{3R^2}\right) = 0.$$

(a) Write down the approximate form of this equation for h small. [§2.8]

(b) Hence write down the design formula for the frequency of small oscillations. [§2.4]

(c) Estimate the frequency of small oscillations of an instrument buoy, radius $R = 0.5$ m, that normally floats half-immersed. (Take $g = 10$ m s^{-2}.)

(This result is of some importance, since the instruments might be sensitive to mechanical resonance at this frequency.)

2.9

(It is important to be able to correlate the gradients and curvatures of a graph with the sign of the differentials u' and u''. Remember, where u' is positive, the curve rises from left to right (falls if u' is negative); and where the second differential u'' is positive, the curve is bent so that the concave face looks upwards (if u'' is negative, the concave face looks downwards).)

Referring to figures 1.3(a)–(f), decide which curves have:

(a) u'' always positive;
(b) u'' always negative;
(c) u' always positive;
(d) u' always negative;
(e) u'' positive or negative and always of opposite sign to u.

2.10

(The following differentials will be needed later in the book.)

(a) Write down the first and second differentials with respect to x of

$$\left(1-\frac{x}{l}\right)\cos \omega x.$$

(Put $1-(x/l)=u$, $\cos \omega x=v$, and use the rule $\mathrm{d}(uv)/\mathrm{d}x=u\,\mathrm{d}v/\mathrm{d}x+v\,\mathrm{d}u/\mathrm{d}x$.)

(b) Find the partial differentials $\partial u/\partial t$, $\partial u/\partial x$, $\partial^2 u/\partial t^2$ and $\partial^2 u/\partial x^2$ for the function

$$u=\cos\left(\frac{x}{l}\right)\sin \omega t.$$

(Remember that $\partial/\partial x$ means differentiate with respect to x treating the other variables as constant.)

(c) Repeat part (b) for the function

$$u=\cos\left(\frac{x}{l}-\omega t\right).$$

Chapter 3

Estimation of time constants by exponential trial functions

3.1 Introduction

Many physical processes involve a smooth change of some variable from an initial to a final value, corresponding to a change of the system from an initial to a final state. For example, the discharge of a capacitor is governed by a simple ODE 1 whose exact solution is

$$Q = Q_0 \, e^{-t/\tau} \tag{3.1}$$

which is illustrated by the familiar exponential decay curve in figure 3.1.

In this chapter, we shall study more complicated equations that describe smooth changes in time. Their solution curves are not exact exponentials, but are sufficiently close to allow exponential TFs to be used to find useful design formulae for the time constant of the change.

3.2 Exponential rise and exponential decay

There are three exponential functions that we shall require for use as trial functions. The first is the *exponential decay function*

$$u = u_0 \, e^{-t/\tau} \tag{3.2}$$

illustrated in figure 3.2 for various values of the *time constant* τ.

The second is the *exponential rise function*

$$u = u_\infty (1 - e^{-t/\tau}) \tag{3.3}$$

shown in figure 3.3.

If a function is required to describe a smooth transition from a nonzero initial state u_0 to a nonzero final state u_∞, then the third function,

$$u = u_\infty + (u_0 - u_\infty) \, e^{-t/\tau} \tag{3.4}$$

may be used. It will be recognised that the functions (3.2) and (3.3) are special cases of the more general *exponential change function* (3.4), having either $u_\infty = 0$ or $u_0 = 0$.

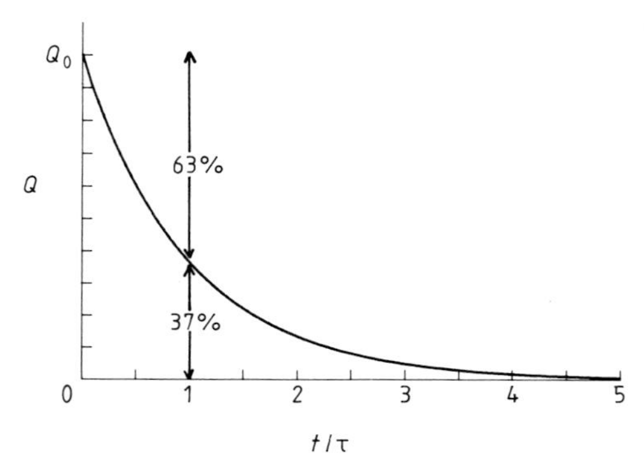

Figure 3.1 Exponential decay of charge Q as a function of time t according to expression (3.1).

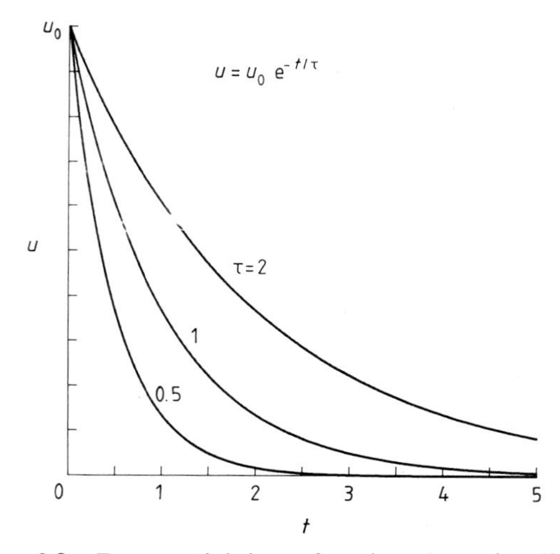

Figure 3.2 Exponential decay functions (equation (3.2)).

A clear physical meaning is attached to every quantity appearing in these functions. The independent variable t denotes time, measured from the start of the process; u_0 and u_∞ are the initial and final values of the dependent variable u, which can represent one of many physical quantities, such as temperature, charge or concentration; finally τ denotes the time constant of the process, which is a measure of the time taken for the system to change from its initial to its final state. Specifically, it is the time for a fraction 0.63 (strictly $1 - e^{-1}$) of the change to be complete.

The notion of a time constant is usefully extended to curves that are not true exponentials but have the same general shape, with an asymptotic approach to a final steady value. We shall still define the time constant as the time to

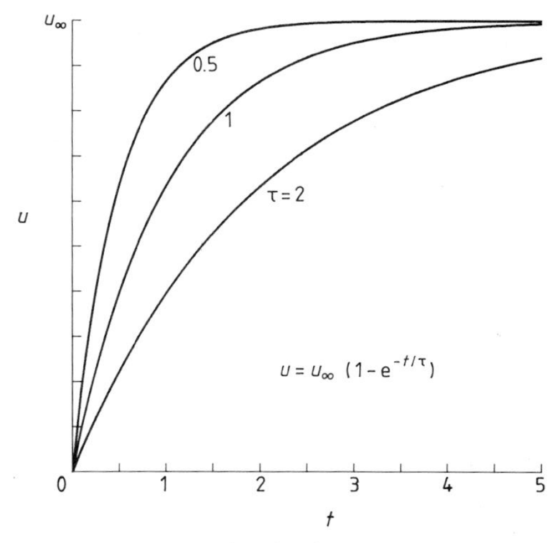

Figure 3.3 Exponential rise functions (equation (3.3)).

complete 0.63 of the whole change. Estimates of τ are then a useful guide to the timescale of the process, whether it is exactly exponential or not. A further discussion of the time constant and the concept of completion is given at the end of this chapter (§3.7). For the present, the simple definition of the time constant is sufficient, and we shall therefore proceed directly to our first example to show how useful estimates of τ may be made for practical problems.

3.3 The falling stone problem

3.3.1 Mathematical model

As explained in Chapter 1, the QSTF method involves:

(a) making the qualitative sketch (QS);
(b) using this to choose a trial function (TF);
(c) using the TF to find a design formula.

Each of these steps will now be explained in detail by considering the velocity v of a stone falling under gravity, retarded by air drag, which obeys the equation

$$m\frac{\mathrm{d}v}{\mathrm{d}t} = mg - bv^2. \tag{3.5}$$

[mass × acceleration = gravity force − air drag]

If the stone starts from rest, $v = 0$ when $t = 0$, so the initial condition is

$$v(0) = 0. \tag{3.6}$$

It is assumed that the stone falls for a sufficiently long time to allow a final steady state to be attained. It is further assumed that the air drag coefficient b remains constant throughout the observation time.

The goal is to understand how the velocity depends on time, and in particular to find a time constant for the process. The reader is asked to ignore for the moment the fact that a relatively simple exact solution exists. This is deliberate, so that in this first example a direct comparison may be made between the results of the approximate trial function method and the exact solution.

3.3.2 Qualitative sketch

The qualitative sketch of the solution curve must be made without formally solving the equation, and there are two distinct means of doing this:

(a) Physical intuition.
(b) Graphical arguments.

Physical intuition is the mental process (largely subconscious) of reviewing all the possibilities and choosing the most reasonable. For the present example, common experience tells us that the velocity of a stone dropped from rest increases steadily from zero at $t=0$ towards a final constant value, the *terminal velocity*. With this statement in mind, the reader is invited to inspect the curves in figure 3.4 and to select the one that most closely matches his expectations.

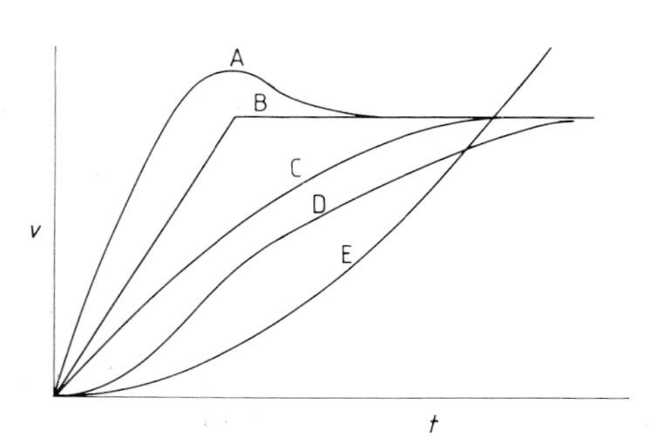

Figure 3.4 Hypothetical sketches of the velocity of a falling stone as a function of time.

Very few people with any scientific background would hesitate to choose curve C, with its smooth approach to a plateau, as the curve closest to the true solution curve. The qualitative sketch can thus be made in this example without further ado. Figure 3.5 is the chosen curve, redrawn to include all the information known already. Notice that physical intuition has allowed us to sketch the solution curve without referring to the equation!

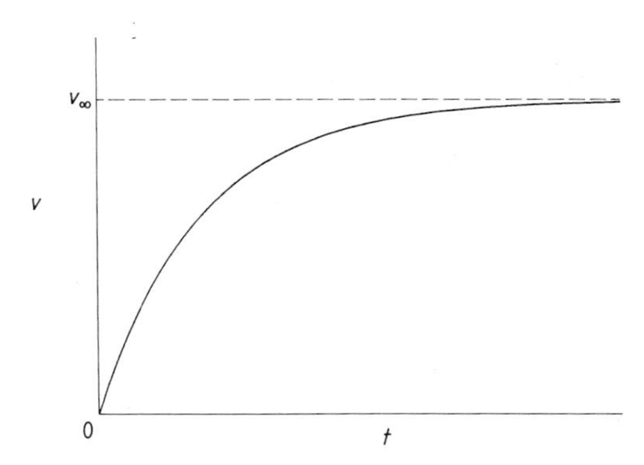

Figure 3.5 Qualitative sketch of $v(t)$ for a falling stone; v_∞ is the terminal velocity.

The alternative graphical argument is as follows:

(a) From the initial condition, $v=0$ when $t=0$.

(b) Putting $v=0$ into equation (3.5), $\mathrm{d}v/\mathrm{d}t=g$, so the initial gradient is positive and v rises as t increases.

(c) This rise in v makes $mg-bv^2$ smaller, so the equation shows that the gradient becomes less steep as time goes on.

(d) Eventually v becomes so large that $mg-bv^2$, and correspondingly $\mathrm{d}v/\mathrm{d}t$, approach zero. The curve therefore flattens off, with the final value of $v=v_\infty$ given by

$$mg-bv_\infty^2=0 \qquad \text{or} \qquad v_\infty=(mg/b)^{1/2}. \tag{3.7}$$

These four facts lead to the same conclusion, that the qualitative sketch must look like figure 3.5. Note that, although the graphical method inspects the equation to study the gradients, there is no question of having to find a formal solution.

3.3.3 Choosing the trial function

Visually the qualitative sketch in figure 3.5 resembles the exponential rise curve (figure 1.3(*c*)), and the next step is to compare the qualitative sketch with more detailed mathematical forms. The three possible exponential functions were described in §3.2, and for this example figure 3.3 and the corresponding function (3.3) are clearly the right choice. Writing v for u as the dependent variable, we therefore choose as the TF

$$v^*=v_\infty(1-\mathrm{e}^{-t/\tau}) \tag{3.8}$$

where v_∞ is given by formula (3.7). We shall always use an asterisk to distinguish a TF.

At this stage we have already acquired a great deal of physical understanding about the equation and its solution without any formal mathematics. We have not 'solved' the equation in the accepted sense; and yet we know that the solution is qualitatively an exponential rise curve, and that the TF (3.8) is the approximate functional solution. The only unknown that remains is the value of the time constant τ; this corresponds to not knowing the timescale for the qualitative sketch.

As explained in §1.2, the purpose of the second stage of the QSTF method is to complete the solution by finding a design formula for τ.

3.3.4 Estimation of the time constant

The systematic technique for finding the unknown constant in any TF involves substituting the TF into the equation and adjusting the unknown constant to achieve the best possible balance. There are two formal steps in this process.

(i) Substitute the TF to form the residual equation
The procedure is to replace v and dv/dt in the governing equation by v^* and dv^*/dt. For equation (3.5) and v^* given by the TF (3.8), this leads to

$$m\frac{v_\infty}{\tau}\mathrm{e}^{-t/\tau}\approx mg-bv_\infty^2(1-\mathrm{e}^{-t/\tau})^2.$$

This can be simplified by replacing bv_∞^2 by mg (equation (3.7)), and dividing by m. The result is

$$\frac{v_\infty}{\tau}\mathrm{e}^{-t/\tau}\approx g-g(1-\mathrm{e}^{-t/\tau})^2.$$

At this stage the two sides only balance approximately because v^* is only an approximate solution. The imbalance can be corrected by adding a term $\mathcal{R}$, called the *residual*, resulting in the *residual equation*

$$\frac{v_\infty}{\tau}\mathrm{e}^{-t/\tau}=g-g(1-\mathrm{e}^{-t/\tau})^2+\mathcal{R}. \tag{3.9}$$

If, by chance, the exact solution had been chosen as the trial function, then it would be possible to make $\mathcal{R}$ zero for all values of t by appropriate choice of τ. Since in this case v^* is only an approximation to the exact solution, $\mathcal{R}$ cannot be made zero everywhere. It is evident that the worse the approximation, the larger $\mathcal{R}$ will be. The residual $\mathcal{R}$ is therefore in some sense a measure of the badness of the approximation.

If equation (3.9) is rearranged as

$$\mathcal{R}=\frac{v_\infty}{\tau}\mathrm{e}^{-t/\tau}-g+g(1-\mathrm{e}^{-t/\tau})^2$$

we can see that $\mathcal{R}$ varies with t and also with the constant τ. The variation is shown graphically in figure 3.6 for three different values of τ. We can therefore generally only make $\mathcal{R}$ zero for one value of t, so we can only balance the

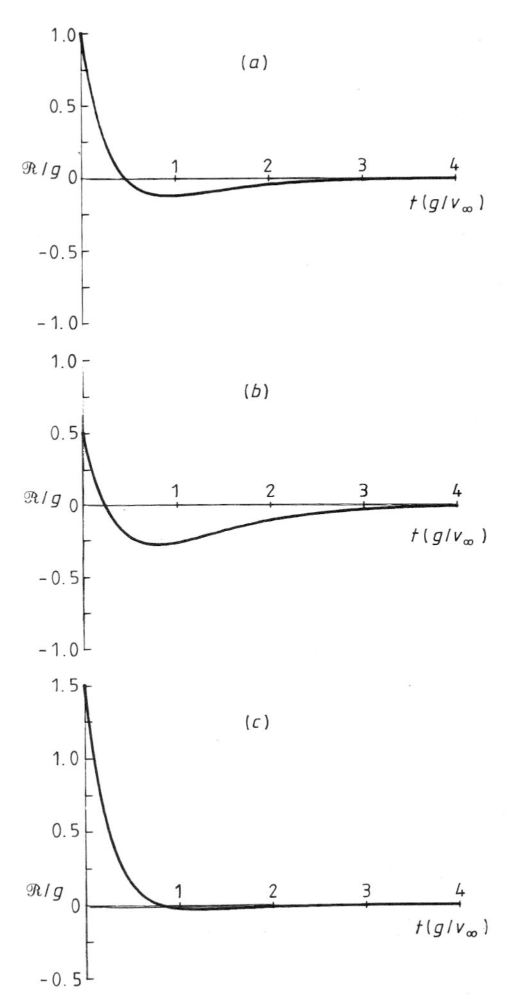

Figure 3.6 Variation of residual $\mathscr{R}$ with t for the falling stone problem for three values of τ: (*a*) $\tau=0.67(v_\infty/g)$; (*b*) $\tau=0.8(v_\infty/g)$; (*c*) $\tau=0.56(v_\infty/g)$. Curve in (*a*) corresponds to halfway collocation.

equation at one point during the change from starting to the final state. The second step is therefore the following one.

(*ii*) *Balance at the halfway stage to find the time constant*
The process of balancing the equation at a selected point is known as *collocation*, and will be considered in more detail in §3.4. For the moment, we shall simply state the rule: for exponential TFs the equation should be balanced halfway through the change. Mathematically, the rule is: put $\mathscr{R}=0$ and

$e^{-t/\tau} = \frac{1}{2}$ in the residual equation. For equation (3.9) the result is

$$\tfrac{1}{2}v_\infty/\tau = g - g(1-\tfrac{1}{2})^2. \tag{3.10}$$

This simple algebraic equation can be solved immediately to give a design formula for the time constant τ. Recalling formula (3.7) for v_∞, the result is

$$\tau = 0.67\left(\frac{m}{gb}\right)^{1/2}. \tag{3.11}$$

3.3.5 Comparison of estimated and exact time constants

It is important to realise what has been achieved in deriving the design formula (3.11). The actual working consists of six lines of simple mathematics:

(1) The original equation (3.5).
(2) The final velocity expression (3.7).
(3) The trial function (3.8).
(4) The residual equation (3.9).
(5) The balanced equation (3.10).
(6) The final design formula (3.11).

The crucial step that simplifies the whole process is balancing at a single point, between lines 4 and 5. The question that naturally arises is how such a dramatically simple step, involving no integration, can achieve a useful result. Repeated experience shows that it does. In this example, for instance, the exact solution for the falling stone equation is

$$v = \left(\frac{mg}{b}\right)^{1/2} \tanh\left[\left(\frac{bg}{m}\right)^{1/2} t\right].$$

To obtain this involves an integration, and the time constant must then be found from the exact formula by setting the tanh function equal to 0.63; the result is

$$\tau = 0.745\left(\frac{m}{gb}\right)^{1/2}. \tag{3.12}$$

It will be seen that the approximate formula (3.11), found by the TF approximation, is numerically well within $\pm 30\%$ of this result. The functional dependence of τ on m, g and b given by the TF result is identical with the exact solution.

The reader can confirm by looking back at this example that the QSTF method described satisfies the requirements for a useful method of approximation discussed in Chapter 1. In particular, it is quick, simple and therefore likely to be free from gross error. The design formula derived by the QSTF method shows the correct functional dependence on the equation parameters, and is accurate to better than $\pm 30\%$. Finally, the result can be easily understood, since it has been sketched graphically and can be described in words: 'v increases from zero as an approximate exponential rise to a final velocity $v_\infty = (mg/b)^{1/2}$, with a time constant $\tau \approx 0.67(m/gb)^{1/2}$.'

3.4 Collocation

Attention was drawn in §3.3.5 to the crucial step of balancing the equation at a single point. The object of this step is to achieve some kind of residual minimisation. Of the several techniques for doing this, the simplest and most widely applicable is collocation, and this is the method used throughout the book for nonoscillatory problems. The following nonrigorous discussion is intended to show that both the procedure and the rule have a rational basis.

The heart of the collocation concept is illustrated in figure 3.7. The notion is that if the curves of two functions believed to be free of violent kinks or discontinuities agree at several points, then one function is a good approximation to the other. A similar assumption is made when a smooth curve is drawn through a set of experimental points. Approximations to a function can therefore be found by choosing a TF with n adjustable constants, and then adjusting these constants so that the TF and the given function agree in value at n points in the domain being investigated. The more of these collocation points, the more accurate the approximation should be but the more complicated the TF and the working. It is also clear that for the best matching the collocation points should be spread out over the whole domain, rather than bunched up in a small region.

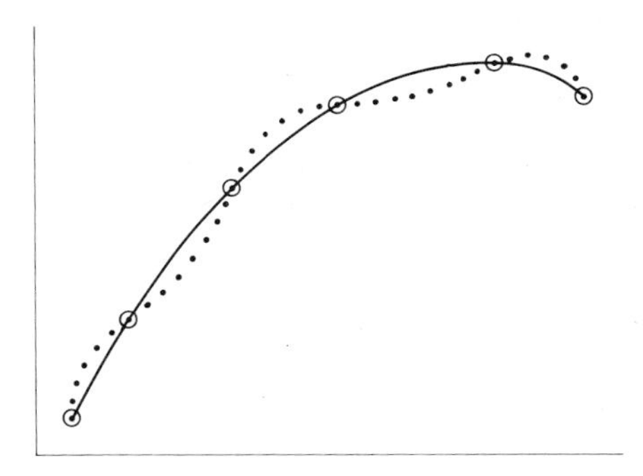

Figure 3.7 Approximation of an exact solution curve (dotted curve) by a simpler curve (full curve) using collocation at six points.

If we now look at the method used in the example, we see that at first the TF was made to satisfy the initial condition and final exact value. This is equivalent to collocation at the endpoints of the curve for the dependent variable v. There is one adjustable constant remaining in the TF, the time constant τ. This can be adjusted to give agreement at one more point. Since collocation has already been performed at $v=0$ and $v=v_\infty$, it seems sensible to collocate halfway between these two values in order to spread the collocation points uniformly across the domain of the dependent variable. Direct collocation with the exact solution is not possible since the latter is unknown except at the ends. The only information available about the solution is embodied in the differential equation, so the expedient is adopted of making an effective collocation by balancing the equation at the halfway stage†.

This qualitative argument does not establish that the most accurate results will be obtained by collocation at exactly halfway. It is possible to devise more elaborate rules for choosing the collocation point to get improved accuracy for a particular equation. It is also possible to improve the accuracy by choosing a TF with more than one adjustable constant and collocating at more intermediate points. This inevitably involves a great deal more labour and gives a less comprehensible result.

Improvements in accuracy, and the relative merits of collocation and other related devices for minimising the residual $\mathscr{R}$, are the field of applied mathematicians and are outside the purposes of this book. The interested reader will find in Crandall (1956) an excellent treatment of these problems from the point of view of the engineer. We must re-emphasise that we are concerned here with the opposite need of providing not increased accuracy and rigour, but increased simplicity and understanding. Experience shows that the simple and easily remembered rule:

'Collocate exponential TFS halfway through the change, i.e. at $e^{-t/\tau}=\frac{1}{2}$'

is entirely adequate for estimating the time constant.

3.5 The bimolecular reaction

3.5.1 Mathematical model

To reinforce the points developed in the preceding sections, we shall consider a second example, the equation governing the kinetics of the bimolecular reaction

$$A+B \rightarrow C.$$

In this reaction, two molecules A and B combine to form a new molecule C. The objective is to find a design formula for the time constant of the process.

The governing equation for the reaction will be taken to be

$$\frac{dn}{dt} = K(a-n) \times (b-n) \tag{3.13}$$

$$\left[\begin{array}{c}\text{rate of formation}\\ \text{of molecules C}\end{array} \propto \begin{array}{c}\text{number of}\\ \text{molecules A left}\end{array} \times \begin{array}{c}\text{number of}\\ \text{molecules B left}\end{array}\right]$$

† The term 'collocation' is being used here in the broad sense of 'securing agreement' at a point. Collocation of the equation is not mathematically identical to collocation of the solutions. However, the difference in the results is unimportant for our purposes.

where n is the number of molecules C after time t and K is a reaction constant. At the start of the reaction ($t=0$), there are no molecules C yet formed, so that the initial condition is

$$n(0)=0.$$

Also in equation (3.13) a and b are constants standing for the numbers of molecules A and B present at the start of the reaction. For every molecule C formed, one molecule A and one molecule B are used up. It is assumed that $a<b$, so the molecules A are used up first.

3.5.2 Qualitative sketch/trial function solution

The solution steps follow the sequence established in §3.3. Now that the explanation of the method has been given, the whole process can be speeded up and presented much more concisely.

(*i*) *Qualitative sketch*
The qualitative sketch shown in figure 3.8 is based on the following facts, which can be arrived at either by using one's physical intuition about chemical reactions, or by inspecting the equations to see how the gradient dn/dt must change.

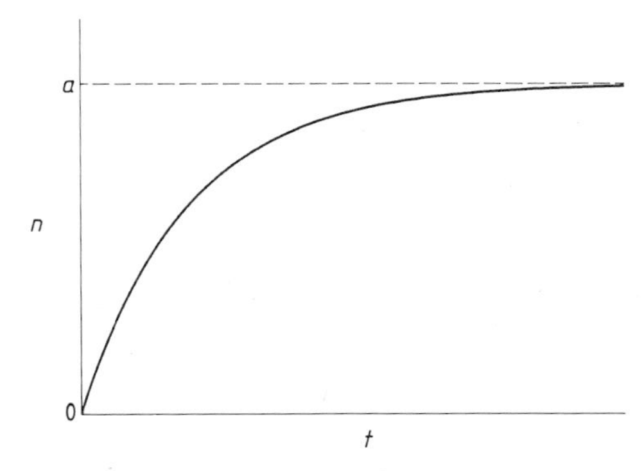

Figure 3.8 Qualitative sketch for the bimolecular reaction.

(a) At $t=0$, no molecules C are yet formed, so $n(0)=0$.

(b) At first there are large numbers of molecules A and B, so the reaction occurs at a high rate. The number of molecules C rises at a high rate, so dn/dt is large and positive.

(c) As the reaction proceeds, molecules A and B are used up and so fewer are available to react. The rate of increase of molecules C, i.e. dn/dt, falls progressively.

(d) Eventually all the molecules A are used up ($a<b$). No further increase in

the number of molecules C can occur, so $dn/dt=0$. Putting this into equation (3.13), the final number of molecules C is $n_\infty = a$.

(ii) Choice of TF

The TF is written down by comparing the qualitative sketch in figure 3.8 with the standard exponential rise curves in figure 3.3. The corresponding function is therefore

$$n^* = a(1 - e^{-t/\tau})$$

where n replaces u and a replaces u_∞ in the standard function (3.3).

(iii) Substitution of the TF *into the governing equation to form the residual equation*

The governing equation is equation (3.13). Substituting the TF gives

$$\frac{a}{\tau} e^{-t/\tau} = K[a - a(1 - e^{-t/\tau})][b - a(1 - e^{-t/\tau})] + \mathscr{R}.$$

(iv) Collocate at $e^{-t/\tau} = \frac{1}{2}$ *to find the time constant* τ

Putting $e^{-t/\tau} = \frac{1}{2}$ and $\mathscr{R} = 0$ gives

$$\frac{a}{2\tau} = K\left(\frac{a}{2}\right)\left(b - \frac{a}{2}\right).$$

Therefore

$$\tau = \frac{1}{Kb[1 - (a/2b)]}. \tag{3.14}$$

This is the required design formula for the time constant. The reader will appreciate just how quick and simple the mathematical derivation of this formula is, now that the steps are presented without the explanatory comments necessarily added to the first example.

3.5.3 Comparison with the exact time constant

Again, this example has an exact solution, which may be written in the form

$$n = \frac{ab(1 - e^{-kt})}{b - a\,e^{-kt}} \qquad (a < b) \tag{3.15}$$

where $k = Kb(1 - a/b)$.

The occurrence of time in both numerator and denominator of this solution makes it difficult to visualise the solution curve or to associate a time constant with it. However, the exact time constant can be deduced numerically by calculating the time for n to reach $0.63a$. The results of doing this for various values of the ratio a/b are shown in figure 3.9. It will be seen that the design

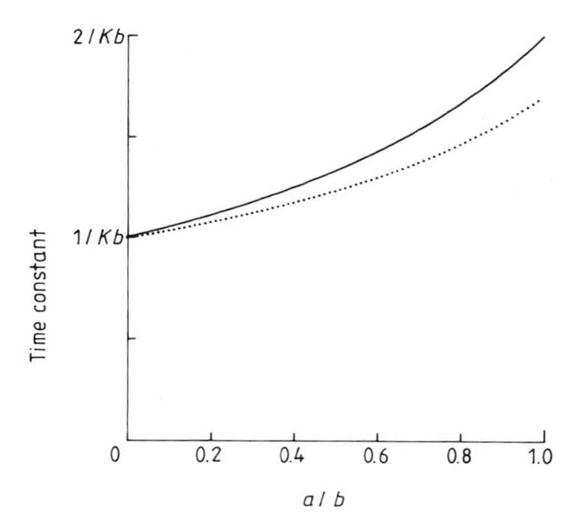

Figure 3.9 Comparison of time constants for the chemical reaction $A+B \rightarrow C$, obtained by the QSTF method (full curve) and from the exact solution (dotted curve). Here a/b is the ratio of the initial numbers of molecules A and B.

formula (3.14) obtained by the TF approximation agrees with the exact value to well within 30% over the whole range of the ratio a/b.

3.6 Checking the trial function hypothesis*

In the two examples so far given to illustrate the use of the TF method, the results have been checked by comparison with the exact results. This provided direct evidence that the method was giving useful results. Often, however, such a comparison is not possible, because an exact solution is not known or it is too complicated to be useful. Indeed, the special utility of the TF method is in just such cases.

It is therefore valuable to have a method of checking that the TF hypothesis is valid, and that the resulting solution or design formula is therefore useful. The method that we now describe requires no knowledge of the solution beyond that available from the TF method itself. It is based on collocation at points other than the halfway stage.

The method will be described by means of a simplified form of the falling stone equation (3.5) in which the constants are all equal to unity. The equation is

$$\frac{\mathrm{d}v}{\mathrm{d}t} = 1 - v^2 \qquad \text{with } v(0) = 0. \tag{3.16}$$

* This section may be omitted at a first reading.

The TF is

$$v^* = 1 - e^{-t/\tau}.$$

Substituting the TF into the equation gives the residual equation

$$\frac{1}{\tau}(e^{-t/\tau}) = 1 - (1 - e^{-t/\tau})^2 + \mathscr{R}.$$

If we collocate at $e^{-t/\tau} = \frac{1}{2}$, we obtain the collocation equation

$$\frac{1}{\tau}(\tfrac{1}{2}) = 1 - (1 - \tfrac{1}{2})^2$$

and the result is $\tau = 0.67$.

If, instead we collocate at $e^{-t/\tau} = 1$, we get

$$\frac{1}{\tau}(1) = 1 - (1 - 1)$$

and the estimate for τ is $\tau = 1$.

Similarly the result for collocation at $e^{-t/\tau} = \frac{1}{5}$ is $\tau = 0.56$. These results are summarised in table 3.1.

Table 3.1 Effect of varying the collocation point on the time constant of a bimolecular reaction.

Collocation point	v^*	$e^{-t/\tau}$	τ
At the beginning	0	1	1
Halfway to completion	$\frac{1}{2}$	$\frac{1}{2}$	0.67
80% complete	$\frac{4}{5}$	$\frac{1}{5}$	0.56

If the true solution of equation (3.16) had been exactly an exponential function, then τ would have remained constant regardless of the collocation point, the equation balancing exactly at every point in the domain. The variation of τ is a sign that the true solution is not an exponential; the further the true solution is from an exponential, the stronger the variation in τ is likely to be. Varying the collocation point is therefore one way of detecting a flaw in the original hypothesis that the exponential is a good approximation to the true solution.

Since collocation is a quick process, it is useful to collocate first at $e^{-t/\tau} = \frac{1}{2}$

and derive the time constant τ in the usual way; and then to collocate again at $e^{-t/\tau}=\frac{1}{5}$. This will give a different value of τ, which we shall denote by τ_c and call the *completion time constant*, for reasons explained in the next section. A good rule is that the results of the exponential trial function method will be sufficiently accurate for design formula purposes provided that

$$\tfrac{1}{2}<\tau/\tau_c<2 \tag{3.17}$$

that is, the ratio of the two time constants does not exceed 2 either way. If they differ by a factor of more than 2, this is a signal that the true solution is too far from an exponential function for the time constants to have any clear meaning, and the results should be discarded or used with caution.

For the falling stone, for instance, $\tau=0.67$ and $\tau_c=0.56$; so the ratio is

$$\tau/\tau_c=1.2.$$

The criterion (3.17) is therefore satisfied.

3.7 Estimation of completion times*

The material in this section is not essential for understanding or applying the method. Its purpose is to extend the reader's understanding of *completion* in exponential or approximately exponential changes, and to enable more accurate numerical estimates to be made of the times required for a process to be completed for practical purposes.

Time constants are often used to estimate *completion times* of exponential processes. Because the curves approach the final value asymptotically, it is not possible to define an exact time when the process can be said to be complete. However, for the true exponential curves, corresponding to the expressions (3.2)–(3.4), it is possible to say what fraction of the change is complete after a time t. We already know that 0.63 of the change is completed when $t=\tau$. By putting $t=2\tau$, 3τ, and so on, in the exponential expressions, we can construct table 3.2, which shows the fraction of change completed after successively longer times.

The engineer thus replaces the idea of completion by *completion for the practical purpose in hand*. In a typical electronic switching circuit, for instance, switching might occur when the capacitor is 80% charged; from table 3.2 this may be seen to require a time between τ and 2τ. A chemical reaction, on the other hand, might not be regarded as complete for practical purposes until 99% of the change has occurred; so a time of about 5τ would have to be allocated. Thus knowledge of the time constant enables the time for practical completion to be worked out, although the multiplying factor varies from one

* This section may be omitted at a first reading.

Table 3.2 Fraction of exponential change complete after various multiples of the time constant τ.

Time elapsed	Fraction of change completed
τ	0.63
2τ	0.86
2.3τ	0.90
3τ	0.95
4τ	0.98
5τ	0.993

application to another. This use of the time constant is well understood in all branches of science and engineering.

If the solution curve is not a true exponential, the time constant τ is still defined as the time to accomplish 0.63 of the whole change, but it is no longer true, for instance, that 0.95 of the change is complete in a time 3τ, since the curve shapes are different in detail. Provided the curves are not too different, however, it is possible to estimate completion times from the TF results by a slight modification to the way table 3.2 is applied. The basis of the method can be seen by studying the way the approximating curves for various collocation points differ from the exact solution. Again, using the falling stone equation (3.16) as an example, figure 3.10(*a*) (broken curve) shows the curve of

$$v = \tanh x$$

which is the exact solution, together with the approximating exponential curves for the three time constants shown in table 3.1 (full curves). Visually, the halfway rule (curve with $\tau = 0.67$) is the best overall choice, but the curve from collocation at the start ($\tau = 1$) gives a better fit close to $t = 0$. Similarly, figure 3.10(*b*), which shows the final stages of these curves on a larger scale, suggests that moving the collocation point nearer completion (collocation at four-fifths of completion, $\tau = 0.56$) gives a much better match for large values of t. The same result is obtained for many other practical examples, namely that collocation at the halfway stage ($e^{-t/\tau} = \frac{1}{2}$) gives a close visual fit for values of t between $\frac{1}{2}\tau$ and 2τ, during which most of the change is accomplished. For the later stages of the change, beyond $t = 2\tau$, a better match is obtained by collocating at $e^{-t/\tau} = \frac{1}{5}$.

It follows that if an estimate of the time for 0.63 completion is required, the standard time constant τ should be used. A better estimate of the times to reach the later stages of completion is obtained by using τ_c. Table 3.3 summarises these points. It should be borne in mind that, if τ_c and τ differ by a factor of more than 2, the exponential TF is a poor approximation to the true solution, and any so-called time constant should be used with great caution (§3.6).

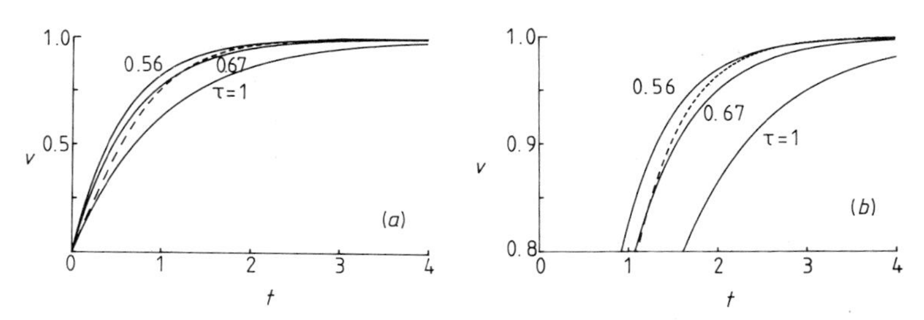

Figure 3.10 (*a*) The exact solution $v = \tanh x$ to equation (3.16) (broken curve), together with the exponential approximation for the time constants listed in table 3.1 (full curves). (*b*) The curves of (*a*) shown with an expanded v scale near completion.

Table 3.3 Estimation of completion times for approximately exponential changes.

Fraction of change completed	Time elapsed	Collocate at
0.63	τ	$e^{-t/\tau} = \frac{1}{2}$
0.86	2τ	
0.95	$3\tau_c$	$e^{-t/\tau} = \frac{1}{5}$
0.98	$4\tau_c$	
0.993	$5\tau_c$	

One other important general point is illustrated by figure 3.10(*a*). The approximating exponential curves found by the QSTF method may not be useful for studying the behaviour of a process near the beginning of the change. In particular, misleading results are often obtained if an attempt is made to calculate the initial slope by differentiation of the TF solution.

Problems

3.1

(a) Confirm that the qualitative sketch of u against t for the equation

$$\frac{du}{dt} = 1 - u^3$$

with initial condition $u(0) = 0$, is similar to the exponential rise curves in figure 3.3. [Use graphical arguments, as §3.3.2]

(b) Write down the final value u_∞. [§3.3.2]

(c) Write down the appropriate TF. [§3.3.3]

(d) Substitute and form the residual equation. [§3.3.4]

(e) Collocate halfway and find a design formula for the time constant. [§3.3.4]

3.2

(a) Make qualitative sketches of u against t and hence write down the appropriate TFS approximating the solutions of the equations

$$\frac{\mathrm{d}u}{\mathrm{d}t}+au^n=0 \qquad u(0)=u_0 \tag{3.18}$$

and

$$\frac{\mathrm{d}u}{\mathrm{d}t}+au^n=1 \qquad u(0)=0. \tag{3.19}$$

[Graphical argument §3.3.2; compare with figures 3.2 and 3.3]

(b) Substitute the TFS and collocate to find design formulae for the time constant τ for the two equations in (a). [§3.3.4]

(c) Confirm that when $n=1$ both these design formulae simplify to formula (2.12) for the linear equations of §2.3.

(These two equations are the nonlinear equivalents of the exponential fall and rise equations (2.8) and (2.10). The results show that τ for the rise is not the same as τ for the fall, except in the special case of $n=1$, when the equations become linear.)

3.3

The rise in temperature of a soldering iron cooled by convection obeys the equation

$$c\frac{\mathrm{d}\theta}{\mathrm{d}t}+h\theta^{5/4}=P$$

where θ is the excess temperature above ambient, c is the heat capacity and h is a convective heat transfer coefficient. The initial condition is that the iron is switched on at $t=0$, when the excess temperature is zero.

(a) Write down the final steady-state condition.

(b) Make a qualitative sketch of θ against t and write down the TF.

(c) Substitute to form the residual equation.

(d) Collocate and find a design formula for the time constant, in terms of c, P and the equilibrium temperature θ_∞. [Eliminate h using the final condition in (a)]

3.4

It was assumed in problem 3.3 that the heat loss from the soldering iron was exclusively caused by convection. In practice, radiation losses are significant, and may predominate, in which case the governing equation is approximately

$$c\frac{\mathrm{d}\theta}{\mathrm{d}t}+a(\theta+300)^4=P.$$

(a) Repeat the calculation of the final condition, and find the design formula for the time constant in terms of c, P and θ_∞. [Eliminate a using the final condition]

(b) Using this result for τ_{rad} for the radiatively cooled iron and the corresponding result τ_{conv} for the convective case (problem 3.3(d)), write down a formula for the ratio of the two time constants τ_{conv}/τ_{rad} in terms of θ_∞.

(c) Find the numerical value of this ratio for the typical case $\theta_\infty=300°$C. (The ratio is close enough to 1 to suggest that a compromise design formula such as

$$\tau=0.8c\theta_\infty/P$$

should be useful in practice, regardless of whether convection or radiation is the dominant cooling process.)

3.5

The design of a particular parachute is such that the drag force is proportional to $v^{1.7}$ throughout the range of speed of normal operation. The governing equation is accordingly

$$\mathrm{m}\frac{\mathrm{d}v}{\mathrm{d}t} \quad + \quad bv^{1.7} \quad = \quad mg.$$

[mass × acceleration + drag force = gravity force]

The initial condition is that the parachute is opened at $t=0$, when the speed of the descent is v_0.

(a) Make a qualitative sketch and write down a suitable TF. [Look at relation (3.4)]

(b) Substitute and collocate at $\mathrm{e}^{-t/\tau}=\frac{1}{2}$. Given that v_∞ is negligible in comparison to v_0, show that the equation after collocation is

$$-\frac{mv_0}{2\tau}+b\left(\frac{v_0}{2}\right)^{1.7}=mg.$$

(c) Noting that $mg=bv_\infty^{1.7}$, and is therefore also negligibly small, find a design formula for τ in terms of m, b and v_0.

3.6

(a) Following the method of §3.5, obtain a design formula for τ for the chemical reaction

$$\mathrm{A}+2\mathrm{B}\rightarrow\mathrm{C}$$

which obeys the equation

$$\frac{dn}{dt} = K(a-n)(b-2n)^2$$

where

$$2a > b.$$

The initial condition is that the number of molecules C is zero at $t=0$.

(b) (May be omitted at a first reading.) By collocating at $e^{-t/\tau} = \frac{1}{5}$ find a design formula for τ_c. [§3.6]

(c) Show that the ratio τ_c/τ for b/a small is $\frac{5}{2}(1+3b/20a)$. [§2.8]

(The result for τ has already been quoted in formula (1.4). The ratio found in (c) suggests that, although the functional form found for τ may be useful, the exponential shape assumed is a poor approximation in this case and that numerical values of τ should therefore be used with caution.)

3.7

The viscosity of liquids that are only available in small quantities can be measured in principle by a capillary viscometer. A fine glass capillary is held vertically and lowered to touch a single drop of the liquid. The rise of liquid up the tube is timed, and the height h in the tube is plotted against elapsed time.

An equation for the variation of height with time is

$$h\frac{d^2h}{dt^2} + \frac{5}{4}\left(\frac{dh}{dt}\right)^2 + bh\frac{dh}{dt} + gh = d.$$

Here $b = 8\nu/R^2$, where ν is the kinematic viscosity to be determined and R is the tube radius; g is the known acceleration due to gravity, and d is a constant proportional to the surface tension.

(Experimental observations show that for viscous liquids the level rises from rest without overshoot to a final steady level, in an exponential-like fashion. The goal is to find a design formula for the viscosity in terms of g, R, the measured time constant τ and the final height h_∞.)

(a) Write down a suitable TF.

(b) Substitute and collocate to obtain the equation

$$\frac{h_\infty^2}{16\tau^2} + \frac{bh_\infty^2}{4\tau} - \frac{1}{2}gh_\infty = 0.$$

(c) Substitute $b = 8\nu/R^2$ and find the design formula for ν. (Note that it is ν that is the goal quantity, and not τ, which is found experimentally.)

(For the usual experimental conditions, when the second term in the result of (c) is negligible,

$$\nu = \frac{gR^2\tau}{4h_\infty}$$

is a useful design formula.)

Chapter 4

Estimation of length constants by exponential trial functions

4.1 Introduction

The *length constant* λ plays the same role as the time constant τ except that, whereas τ is used to define a timescale for an exponential change, λ is used to define a distance scale.

Equations involving changes in the dependent variable with distance are most commonly ODE 2, and the two auxiliary conditions needed for a complete solution are almost always boundary conditions. Consequently, whereas time-constant problems are usually associated with ODE 1, subject to an initial condition specified at $t=0$, length-constant problems usually involve the study of ODE 2, subject to boundary conditions.

4.2 The length constant of a long cooling fin

4.2.1 Mathematical model

Figure 4.1 shows a cooling fin of thin rectangular section projecting from a hot plate held at a fixed temperature. The fin loses heat by convection to the surrounding air, and the governing equation, found by expressing the heat balance in a small length element δx, is

$$Kab\frac{d^2\theta}{dx^2} \quad - \quad 2ah\theta^n \quad = 0. \tag{4.1}$$

$$\left[\begin{matrix}\text{rate of heat}\\ \text{gained by conduction}\end{matrix} - \begin{matrix}\text{rate of heat}\\ \text{lost by convection}\end{matrix} = 0\right]$$

Here θ is the excess temperature above that of the surrounding air at a distance x along the fin, K is the thermal conductivity (assumed constant), a is the height of the fin, b is its thickness ($b \ll a$) and h is a convective heat transfer coefficient. The index n may vary from 1 to $\frac{5}{4}$ according to the conditions in the surrounding air.

The equation may be simplified by dividing throughout by Kab. It may then

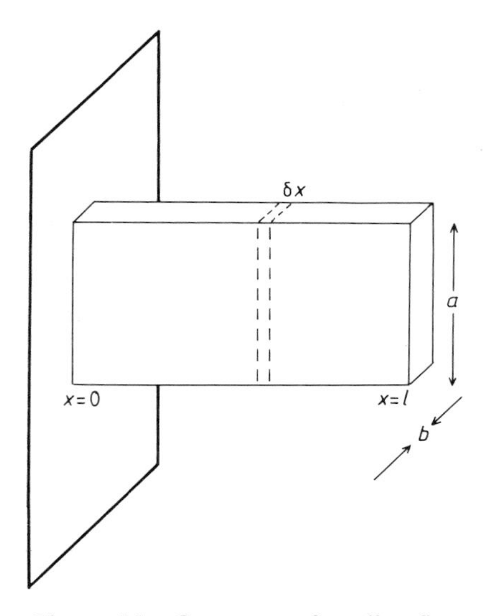

Figure 4.1 Geometry of cooling fin.

be written as

$$\frac{d^2\theta}{dx^2} - c\theta^n = 0 \tag{4.2}$$

where

$$c = 2h/Kb. \tag{4.3}$$

Equation (4.2) is analogous to the linear equation (2.13), but is nonlinear and cannot be solved exactly.

Unlike the examples in the previous chapter, where the independent variable (the time t) had no upper limit, in this example there are two boundaries; the boundary at $x=0$ we shall call the *near boundary*, while the other end of the fin at $x=l$ we shall call the *far boundary*.

The near boundary condition is given by the fixed temperature θ_0 of the base of the fin at $x=0$, i.e.

$$\theta(0) = \theta_0. \tag{4.4}$$

The behaviour at the far boundary depends on the length of the fin and on the details of the heat loss from the end. However, if the fin is made sufficiently long, it is clear that the excess temperature must eventually approach zero, and remain approximately zero along the remote parts of the fin. For long fins, the far boundary condition is therefore $\theta=0$.

The goal of the problem is to find a length constant that characterises the fall in temperature along the fin.

4.2.2 *Qualitative sketch/trial function solution*

(i) Qualitative sketch
Figure 4.2 is based on the following information:

(a) The excess temperature at $x=0$ is θ_0 (near boundary condition).
(b) For a very long fin, the excess temperature at the far end is zero (far boundary condition).
(c) The excess temperature remains zero for some distance, so there is a plateau at the far end of the curve.
(d) The temperature must fall steadily from the base at $x=0$ and approach the plateau asymptotically.

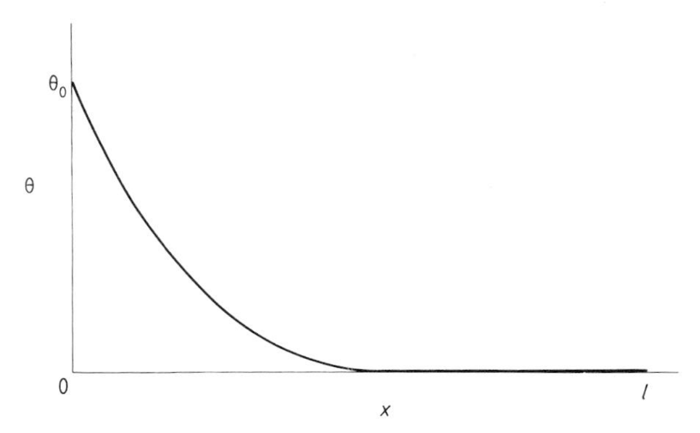

Figure 4.2 Qualitative sketch for the temperature along a long cooling fin.

(ii) Choice of TF
Comparison of the qualitative sketch with curves of the exponential functions shows that it corresponds to figure 3.2. The mathematical form of this decay function is

$$\theta^* = \theta_0 \, e^{-x/\lambda} \tag{4.5}$$

where x and λ are used instead of t and τ of the previous chapter.

Note that this TF satisfies the near boundary condition (4.4). At the far boundary $(x=l)$

$$\theta^*(l) = \theta_0 \, e^{-l/\lambda}$$

and also

$$\theta^{*\prime}(l) = \frac{-\theta_0}{\lambda} \, e^{-l/\lambda}.$$

The assumption is that l/λ is sufficiently large for both θ^* and $\theta^{*\prime}$ to be close to zero for some distance, corresponding to the plateau in figure 4.2. The condition for this assumption to be justified will be discussed in §4.4.

(iii) Substitution and collocation to find λ

Substituting the TF (4.5) into equation (4.2) leads to the residual equation

$$\frac{1}{\lambda^2}\theta_0\,e^{-x/\lambda}-c\theta_0^n(e^{-x/\lambda})^n=\mathscr{R}. \tag{4.6}$$

Collocation at $e^{-x/\lambda}=\frac{1}{2}$ gives

$$\frac{1}{\lambda^2}\frac{\theta_0}{2}-c\theta_0^n(\tfrac{1}{2})^n=0.$$

Solving for λ, we find

$$\lambda=\left(\frac{2}{\theta_0}\right)^{(n-1)/2}c^{-1/2}. \tag{4.7}$$

This is the required design formula for the length constant.

4.2.3 *Understanding the solution*

If the hypothesis of the exponential TF (4.5) is valid, then the temperature along the fin is appproximately

$$\theta=\theta_0\,e^{-x/\lambda}$$

where λ is given by the expression (4.7). Although λ appears to be a rather complicated function of the system parameters, the effect of n and θ_0 is quite small for practical values.

Since n varies between 1 and $\frac{5}{4}$ for most convection conditions, the index $\frac{1}{2}(n-1)$ always lies between 0 and $\frac{1}{8}$. The excess temperature θ_0 is not usually greater than 500 K, so the values of λ given by formula (4.7) lie in the relatively narrow range between $c^{-1/2}$ and $0.5c^{-1/2}$. The importance of the TF result is that it shows that the formula for λ is not greatly dependent upon the exact convection law assumed. This suggests that the assumption often made without justification in 'exact' mathematical investigations, i.e. that $n=1$, will not cause gross errors even if n is nearer to $\frac{5}{4}$.

4.3 The length constant of a filament lamp: the effect of symmetry

4.3.1 *Mathematical model*

Figure 4.3 shows a form of simple filament lamp that is symmetrical about the midpoint. The filament is in the form of a straight wire of length L and diameter D, supported at either end by thick conductors at temperature T_0.

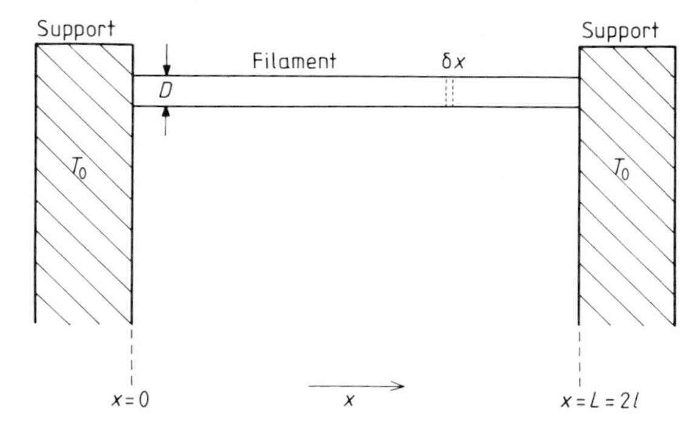

Figure 4.3 Schematic diagram of a simple filament lamp. The small volume element δV is given by $\delta V = \frac{1}{4}\pi D^2\, \delta x$.

The governing equation, based on the balance between the rate of heat losses and gains for a small volume element δV, is

$$-K\frac{d^2T}{dx^2} + \frac{4\varepsilon\sigma}{D}T^4 = \frac{16I^2\rho}{\pi^2 D^4} + \frac{4\varepsilon\sigma}{D}T_a^4.$$

$$\left[\begin{matrix}\text{heat lost by}\\ \text{conduction}\end{matrix} + \begin{matrix}\text{heat lost by}\\ \text{radiation}\end{matrix} = \begin{matrix}\text{heat}\\ \text{generated}\end{matrix} + \begin{matrix}\text{heat gained by}\\ \text{radiation from background}\end{matrix}\right] \tag{4.8}$$

In this equation, K is the thermal conductivity, ε is the emissivity, σ is the Stefan–Boltzmann constant, T is the absolute temperature at a distance x along the filament, I is the current through the filament, ρ is the electrical resistivity and T_a is an effective background radiation temperature. Factors K, ε and ρ are taken to be constants, which is a rather drastic approximation in the model.

Since our present purpose is to illustrate the concept of a length constant for symmetrical systems, it will be sufficient to consider the simplest version of the problem. The end temperature T_0 will be taken as zero, and the heat gained by radiation from the background will be neglected. Equation (4.8) then becomes

$$-\frac{d^2T}{dx^2} + AT^4 = B \tag{4.9}$$

where

$$A = 4\varepsilon\sigma/DK \tag{4.10}$$

and

$$B = 16I^2\rho/\pi^2 D^4 K. \tag{4.11}$$

The boundary conditions are

$$T(0)=0 \qquad T(L)=0. \tag{4.12}$$

The goal is to find a length constant that characterises the temperature change along the filament.

4.3.2 Qualitative sketch/trial function solution

(i) Qualitative sketch

The qualitative sketch (figure 4.4(*a*)) is made by using the following information:

(a) $T=0$ at the two ends (conditions (4.12)).

(b) For a very long filament, the central portion is not much influenced by the cooling effect of the end supports, and the temperature will be substantially constant.

(c) The temperature falls smoothly at each end because of conduction to the supports.

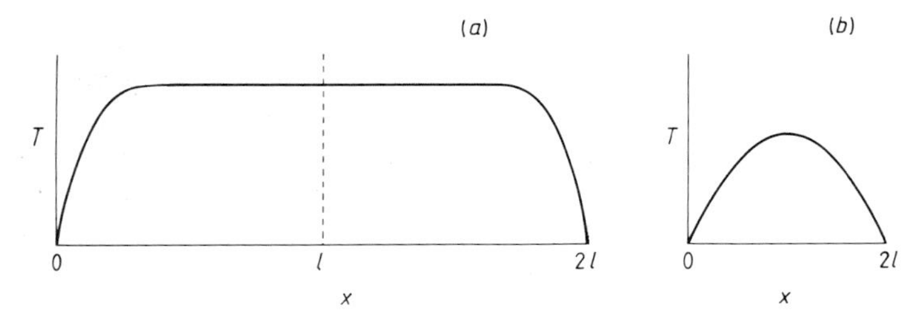

Figure 4.4 (*a*) Qualitative sketch for a long filament lamp showing the plateau and symmetry about the midpoint. (*b*) Sketch for a much shorter filament with no plateau.

As in the previous example, it is assumed that the length L is sufficiently great for a well developed plateau to appear in the solution curve†.

A quite new feature of the sketch is the *symmetry* about the midpoint. The most important consequence of this is that the problem may be more easily solved by folding the solution curve back on itself so that only the left half of the solution is required. For this reason, it is convenient to write the length as $2l$ instead of L. The solution curve for the left half (figure 4.5) now resembles the familiar exponential rise with boundaries at $x=0$ and $x=l$.

† For a much shorter filament (figure 4.4(*b*)), the end effects will be felt right to the centre. There will then be no plateau, and the cooling of the ends will reduce the central temperature significantly.

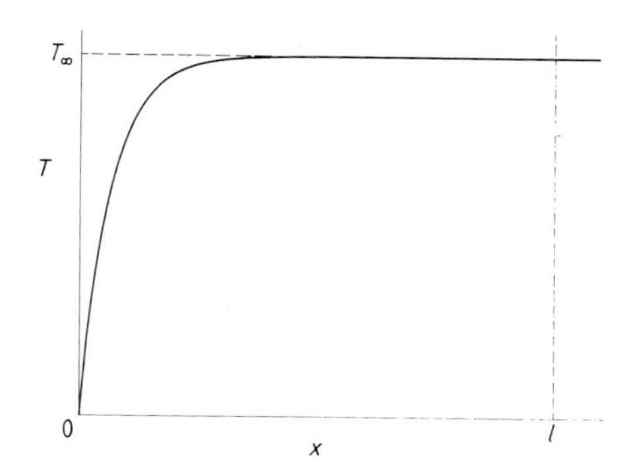

Figure 4.5 Qualitative sketch for the filament lamp (figure 4.4(*a*)). The real boundary at $x = L$ is replaced by a 'symmetry boundary' at $x = l = \frac{1}{2}L$.

(ii) Choice of TF

Comparison of the qualitative sketch (figure 4.5) with the standard exponential curves and their functions (§3.2) suggests the TF

$$T^* = T_\infty(1 - e^{-x/\lambda}). \tag{4.13}$$

The plateau temperature T_∞ may be deduced from the governing equation in a similar manner to that used to find the terminal velocity of the falling stone. For the plateau of an exponential curve, the gradient is practically constant and the second differential (the rate of change of gradient) is negligibly small. We therefore put

$$d^2T/dx^2 = 0$$

into the equation. It then follows that

$$T_\infty = (B/A)^{1/4}. \tag{4.14}$$

(iii) Substitution and collocation to find the length constant

Substituting the TF (4.13) into equation (4.9), we obtain

$$\frac{T_\infty}{\lambda^2} e^{-x/\lambda} + AT_\infty^4(1 - e^{-x/\lambda})^4 = B + \mathscr{R}. \tag{4.15}$$

Collocation at $e^{-x/\lambda} = \frac{1}{2}$ gives

$$\frac{T_\infty}{2\lambda^2} + \frac{AT_\infty^4}{16} = B. \tag{4.16}$$

Combining equations (4.14) and (4.16), we obtain

$$\lambda = 0.73(AB^3)^{-1/8}. \tag{4.17}$$

The original system parameters may now be restored by using the relations

(4.10) and (4.11), whence

$$\lambda = 0.22 \left(\frac{D^{13} K^4}{\varepsilon \rho^3 \sigma I^6} \right)^{1/8}. \tag{4.18}$$

This is the required design formula for λ.

We infer that the temperature along a long heated filament will rise approximately exponentially from either end towards the centre, with a length constant λ given by formula (4.18). If $l \gg \lambda$, the temperature will be nearly constant along most of the filament; the value of T_∞ is given by combining formulae (4.14), (4.10) and (4.11).

4.3.3 *Application to the design of photometric lamps**

As an illustration of the usefulness of design formulae for length contents, we discuss here the design of photometric lamps.

Photometric lamps are designed with a straight filament of length $2l$, rigidly supported at both ends. To be useful for photometry, it is essential that the temperature be substantially constant over the central part of the filament, taken here to be from $x = l/2$ to $x = 3l/2$, as shown in figure 4.6. To be specific, let it be supposed that the temperature is to be constant to within 1% for this central part. The goal of the problem is to find a design formula for the shortest length that will achieve this.

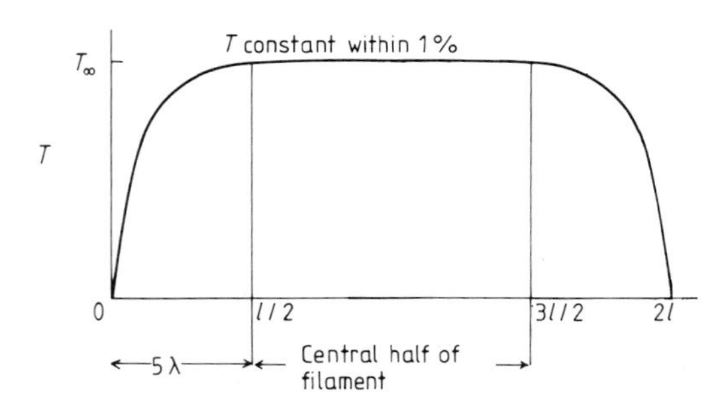

Figure 4.6 Design geometry of a photometric lamp. The central plateau from $x = l/2$ to $x = 3l/2$ is flat to within 1%.

It is clear that in this central region the temperature changes must be nearly complete, so that the plateau can be as flat as possible. For an exponential, table 3.2 shows that 99% completion is attained when $x = 5\lambda$. It follows that at a distance $x = 5\lambda$ from each end, the temperature is within 1% of the completion temperature. The distance $x = 5\lambda$ must therefore be made to

* This section may be omitted at a first reading.

correspond to a quarter of the whole length (see figure 4.6), and hence the minimum length to meet the goal is 20λ.

The design formula for the minimum length of filament for a photometric lamp is therefore 20 times the value of λ given by the design formula (4.18), or

$$\text{minimum length} = 4.4\left(\frac{D^{13}K^4}{\varepsilon\rho^3\sigma I^6}\right)^{1/8}. \tag{4.19}$$

(In practice, the lamps are made with a filament of rectangular, rather than circular, cross-section and λ_c rather than λ should be used (§3.7). The constants in the equation are slightly altered but the method of solution is identical, and the final design formula is essentially the same.)

4.4 Long-range problems

The two examples just given illustrate the use of exponential trial functions in problems involving a specific range for the independent variable x, set by the physical boundaries of the system studied. For the cooling fin problem (§4.2), the base of the fin is at $x=0$ and the free end is at $x=l$. For the filament lamp, the ends of the filament are at $x=0$ and $x=L=2l$, but symmetry allows the far boundary to be located at the midpoint $x=l$. In both examples, therefore, the independent variable ranges from 0 to l. We thus call l the *range* of x or simply the *range*.

It is clear that exponential TFs are a suitable choice when the qualitative sketch shows a well developed plateau (see figures 4.2 and 4.5). We have already seen that such a plateau exists when $l \gg \lambda$, for it then follows that

$$e^{-l/\lambda} \ll 1$$

and this means that the physical change described by the exponential TF is complete for practical purposes well within the range $0<x<l$.

The lower limit to the range for the exponential TF to be satisfactory can be inferred from figure 4.7. Figure 4.7(a) shows that, when $l=6\lambda$, the plateau develops before the curve is cut off by the far boundary: however, if l is reduced to 2λ, the curve is cut off before it reaches the plateau (figure 4.7(b)). The useful lower limit evidently lies between these two values. The working rule suggested by this argument, and confirmed by experience, is to use an exponential TF if the range is more than four length constants, that is, if

$$l \geqslant 4\lambda. \tag{4.20}$$

A problem in which this condition is satisfied will be called a *long-range problem*. If the range is less than 4λ, an exponential TF should not normally be used. The TFs that are most useful for tackling such cases are parabolas, which will be discussed in detail in Chapter 5.

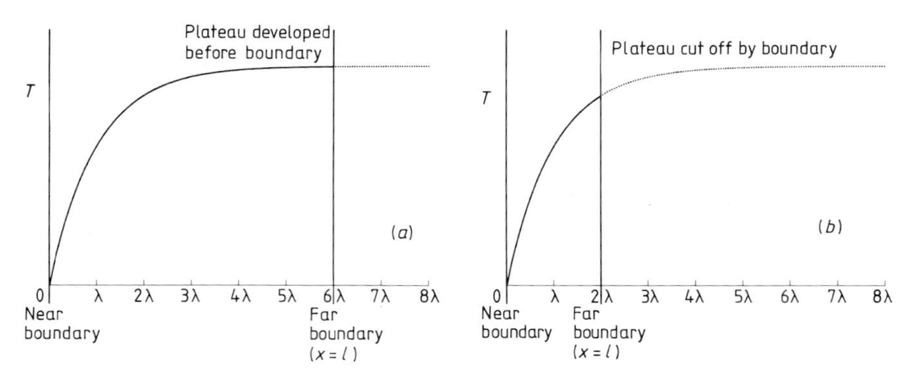

Figure 4.7 Curves showing how the characteristic plateau of an exponential is cut off by the far boundary, when the distance l to the far boundary is reduced from 6λ to 2λ.

In some cases, it will not be apparent at the outset whether a problem falls into the long-range category. This does not matter; the procedure in such cases is to start by assuming an exponential TF, and to use it to deduce a length constant. The 4λ rule may then be applied retrospectively; if it turns out that $4\lambda > l$, then the exponential TF hypothesis is unsuitable and the problem should be reworked using a parabolic TF, as explained in Chapter 5.

Problems

4.1

For practical purposes, the change of speed with distance x descended by a parachute is more interesting than the change of speed with time, which was studied in problem 3.5.

Noting that

$$\frac{dv}{dt} = \frac{dx}{dt}\frac{dv}{dx} = v\frac{dv}{dx}$$

show that the equation of problem 3.5 can be transformed to the ODE 1

$$mv\frac{dv}{dx} + bv^n = mg = bv_\infty^n$$

where the drag force is taken to be proportional to v^n; the equation is thus more general than in problem 3.5, where $n = 1.7$.

(a) Make a qualitative sketch of v against x, and write down the TF.

(b) Write down the residual equation, using the fact that, since $v_\infty \ll v_0$, all terms in v_∞ can be neglected.

(c) Collocate, and derive a design formula for the length constant of the descent.

(The result shows that, for $n=2$, λ is independent of initial velocity and that, if parachutes are designed with $n>2$, the length constant will *decrease* as the initial velocity is increased.)

4.2

If a flat plate is placed edge-on into a stream of moving fluid, the velocity of the stream is undisturbed a long way from the plate, but at the surface of the plate it is reduced to zero, as shown in figure 4.8. The equation relating the dimensionless velocity u to the dimensionless distance x from the surface, known as the *Blasius equation*, is

$$\frac{d^2u}{dx^2}+\left(\int_0^x u\,dx\right)\frac{du}{dx}=0$$

with boundary conditions

$$u(0)=0 \qquad u(\infty)=1.$$

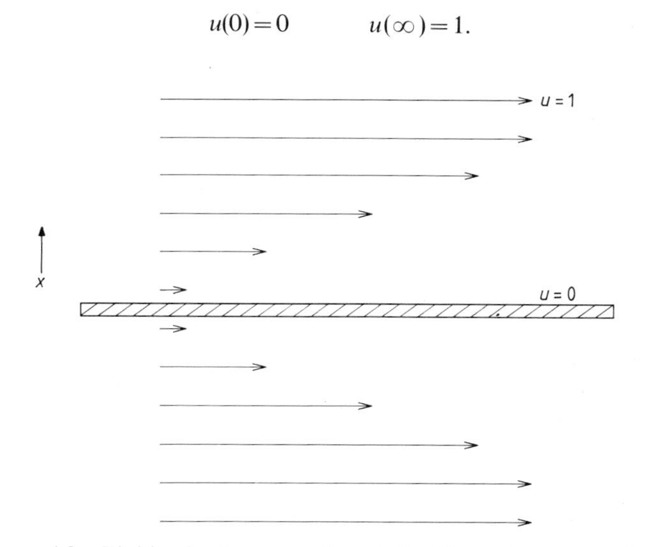

Figure 4.8 Fluid velocity near a thin plate. The vector arrows indicate the magnitude of u. The plate is located at $x=0$.

The physical picture suggests that the qualitative sketch is an exponential rise, with TF

$$u^*=1-e^{-x/\lambda}.$$

(a) Find the expression for $\int_0^x u^*\,dx$ by integrating the TF.
(b) Substitute the TF and this integral to obtain the residual equation.
(c) Collocate at $e^{-x/\lambda}=\frac{1}{2}$ (where $x=\lambda \ln 2$) to find an estimate for the value of λ.

(This problem illustrates the way in which the TF method can be extended to *integro-differential equations*. The value of λ obtained is useful for checking that no gross error has been made in the exact numerical solution, which gives $\lambda = 2.0$.)

(d) The most important numerical quantity required from the model is the gradient of u at the plate. Differentiate the TF and use the value of λ obtained in (c) to estimate the value of du/dx at $x=0$.

(The correct computed answer is $u'(0) = 0.332$. This illustrates the danger of differentiating an approximate solution, which in this case has led to an exaggeration of the error to 30%, which is barely acceptable.)

4.3

The governing equation for the excess temperature θ above ambient of an electric fuse wire of length $2l$ (see §5.2) is

$$-\frac{d^2\theta}{dx^2} + a\theta = b.$$

The boundary conditions are

$$\theta(0) = 0 \qquad \theta(2l) = (0).$$

(a) Make a qualitative sketch of the possible temperature variations with distance for a long and a short fuse; hence write down a suitable exponential TF for the long-range problem. [§4.4]

(b) Noting that the equation is a linear ODE 2 with constant coefficients, write down the design formula for the length constant. [§2.4]

(c) Use this formula for the length constant to find a formula for the minimum length of fuse for which the exponential TF is likely to be a reasonable approximation to the qualitative sketch. [§4.4]

4.4

(May be omitted at a first reading.) By collocating equation (4.15) at $e^{-x/\lambda} = \frac{1}{5}$, show that the 'completion length constant' is given by the same design formula as (4.18), except that the numerical constant is 0.17. [See §§3.6 and 3.7 for concept of completion constant]

Chapter 5

Parabolic trial functions

5.1 Introduction

The solution curves discussed in this chapter resemble all or part of one of the four symmetrical curves shown in figure 5.1. These curves are described by the following *parabolic trial functions*:

figure 5.1(*a*)

$$u^* = u_0 + (u_e - u_0)\left[1 - \left(\frac{x}{l}\right)^2\right] \tag{5.1a}$$

figure 5.1(*b*)

$$u^* = u_0 + (u_e - u_0)\left[1 - \left(\frac{x-l}{l}\right)^2\right]. \tag{5.1b}$$

All the curves in figure 5.1 have the same shape. Each upper curve differs from the corresponding lower one only in the sign of $(u_e - u_0)$. The left-hand curves differ from the right-hand curves only in the choice of origin; replacing x by $(x-l)$ in equation (5.1a) leads to equation (5.1b).

The application of parabolic TFS to ordinary differential equations is very similar to the approach established for exponential TFS in Chapters 3 and 4. One difference is that, whereas the unknown constant in exponential TFS is always the time constant (or length constant), in a parabolic TF there are two common possibilities†:

(a) The boundaries are specified and therefore the constant l in the TFS (5.1) is known; the unknown quantity is then the amplitude $(u_e - u_0)$, which defines the vertical scale, and the goal is a design formula for this quantity.

(b) Two initial conditions are given, which fix the values of u_0 and u_e. The goal is then to find a design formula for the constant l which fixes the horizontal scale. Physically, l usually corresponds to a time constant or period, since for problems in which initial conditions are specified the independent variable is usually time.

† A third possibility occurs in connection with eigenvalue equations (see §9.3).

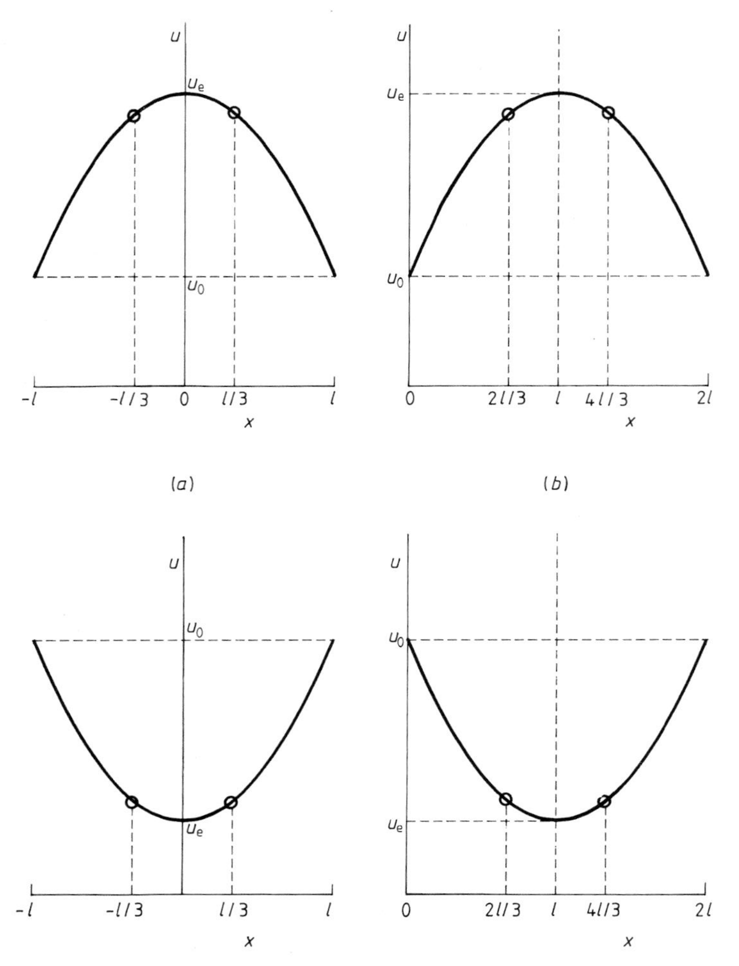

Figure 5.1 Parabolic curves to be used in conjunction with the parabolic TFs given in equation (5.1). The small open circles indicate the normal collocation points.

Apart from the choice of function the only other change in the TF procedure is the rule for collocation. Instead of dividing the vertical scale and collocating halfway up, we divide the horizontal scale and collocate one-third of the way across. More precisely, the most useful rule for parabolic TFs is:

> 'Collocate at a point one-third of the way
> from the extremum to the boundary.'

Collocation points chosen according to this rule are marked by circles in figure 5.1(*a*) and (*b*). Although it might appear that the rule suggests two alternative collocation points, the examples will show that this causes no ambiguity. The 'one-third of the way across' rule is discussed further in §5.3.

5.2 The short fuse

As a first practical application of the parabolic TF, we shall consider an electric fuse, which consists of a relatively thin wire of length L stretched between two stout supports. When current passes through the wire it is heated, and when the current reaches a certain value (the fuse's 'current rating') the wire melts, opening the circuit and preventing damage to other components. The governing equation for the power balance in a small volume δv will be taken as

$$-K\frac{d^2\theta}{dx^2} \quad + \quad \frac{4h\theta}{D} \quad = \quad \frac{16I^2\rho}{\pi^2 D^4}$$

$$\left[\begin{matrix}\text{rate of heat loss from} \\ \delta v \text{ by conduction}\end{matrix} + \begin{matrix}\text{rate of heat loss from} \\ \delta v \text{ by convection}\end{matrix} = \begin{matrix}\text{rate of heat} \\ \text{generated in } \delta v\end{matrix}\right]$$

or

$$-\frac{d^2\theta}{dx^2}+a\theta=b \tag{5.2}$$

where

$$a=\frac{4h}{KD} \tag{5.3a}$$

and

$$b=\frac{16I^2\rho}{\pi^2 D^4 K}. \tag{5.3b}$$

Here θ is the excess temperature above that of the supports and the ambient air, for both of which $\theta=0$. K is the thermal conductivity, D is the wire diameter, h is a convective heat transfer coefficient, I is the current and ρ is the electrical resistivity. Convection has been assumed to be proportional to θ, and radiation has been neglected (see problem 5.1 for alternative assumptions).

The boundary conditions are

$$\theta(0)=0 \qquad \text{at near boundary } x=0 \tag{5.4a}$$

and

$$\theta(L)=0 \qquad \text{at far boundary } x=L. \tag{5.4b}$$

The goal is to find a design formula for the maximum temperature θ_c at the centre of the fuse.

The qualitative sketch is drawn using the same arguments that were used for

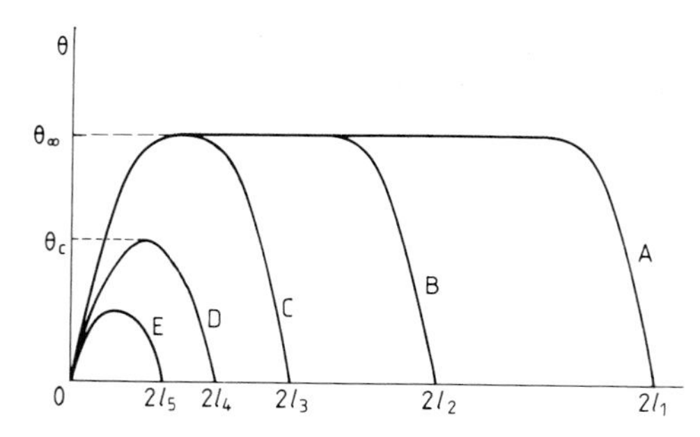

Figure 5.2 Qualitative sketch for a fuse, showing temperature θ against distance x for various lengths $2l$.

the filament lamp (§4.3). Figure 5.2 (curve A) shows the sketch for a very long fuse, with the long central plateau, the supports at zero excess temperature and smooth curves falling from the plateau to the ends.

Figure 5.2 also shows how the curve must change as the fuse is made progressively shorter, starting from the very long case just considered (curve A). At first, the central plateau becomes shorter (curve B) and eventually it must disappear (curve C), the two end-effect curves just meeting in the centre. Any further shortening of the wire (curves D and E) must result in a lowering of the central temperature. Physically, this is a result of the cooling effect of the supports now being felt at the centre of such a short wire. Mathematically, the effect can be studied by rewriting equation (5.2) to show how the temperature depends on the curvature, which is proportional to θ''. Thus at the centre

$$\theta_c = \frac{b + \theta_c''}{a}. \tag{5.5a}$$

For very long fuses, the curvature at the centre is substantially zero, and the temperature is given by putting $\theta_c''=0$ to get

$$\theta_\infty = b/a. \tag{5.5b}$$

For shorter fuses, as in curve D of figure 5.2, the curvature at the centre is obviously not zero; the 'concave downward' curvature corresponds to a negative value of θ_c'' (see problem 2.9). The effect is to lower the value of the RHS of equation (5.5a), and hence to lower the central temperature.

The qualitative sketch shows that there are two distinct limiting problems. For a long fuse, the θ/x curve has a long central plateau, and each half looks like an exponential. This is the long-range problem ($l/\lambda > 4$), which has already been solved using an exponential TF in problem 4.3. For a short fuse, on the other hand, the qualitative sketch is like curves D or E in figure 5.2, and the θ/x curve has no plateau. We shall call such cases *short-range problems*.

The qualitative sketch (e.g. curve D in figure 5.2) suggests a parabola as a TF, and since the x domain extends from 0 to $L=2l$, the matching standard curve is figure 5.1(*b*) (upper), with θ replacing u. With $u_0=0$ and $u_e=\theta_c$ the TF that follows from the corresponding expression (5.1b) is

$$\theta^* = \theta_c\left[1-\left(\frac{x-l}{l}\right)^2\right]. \tag{5.6}$$

We shall assume that the length l is given, and that the temperature at the centre, θ_c, is the unknown goal parameter. The goal is a design formula for θ_c as a function of l.

5.2.1 Solution

Substituting the TF into the governing equation (5.2) leads to the residual equation

$$-\left[-\theta_c\left(\frac{2}{l^2}\right)\right]+a\theta_c\left[1-\left(\frac{x-l}{l}\right)^2\right]=b+\mathcal{R}.$$

Collocation 'one-third of the way across' requires that $\mathcal{R}=0$ at $x=\frac{2}{3}l$ or $\frac{4}{3}l$ (see figure 5.1(*b*)). In either case

$$\left(\frac{x-l}{l}\right)^2=\frac{1}{9}$$

so collocation gives

$$\frac{2\theta_c}{l^2}+\frac{8}{9}a\theta_c=b. \tag{5.7}$$

Solving for θ_c gives the required design formula:

$$\theta_c=\frac{b}{2/l^2+\frac{8}{9}a}. \tag{5.8}$$

This formula can be written in the dimensionless form

$$\frac{\theta_c}{\theta_\infty}=\frac{al^2}{2+\frac{8}{9}al^2}$$

where θ_∞ is given by formula (5.5b). A final substitution is to write $a=1/\lambda^2$, where λ is the length constant derived for a long fuse (see problem 4.3). This gives the design formula

$$\frac{\theta_c}{\theta_\infty}=\frac{(l/\lambda)^2}{2+\frac{8}{9}(l/\lambda)^2}. \tag{5.9}$$

5.2.2 Discussion

This first example was deliberately chosen to have an exact solution, from which the exact formula

$$\frac{\theta_c}{\theta_\infty} = 1 - \frac{2\,e^{-l/\lambda}}{1 + e^{-2l/\lambda}} \tag{5.10}$$

can be derived.

The approximate formula (5.9) has the advantage of a clearer physical content. It shows that for small l the central temperature θ_c increases as l^2, but that eventually the convection losses (second term in denominator) become more significant and prevent the temperature from rising indefinitely.

Figure 5.3 shows the curves of the design formula (5.9) and the exact formula (5.10). For short fuses ($l < 4\lambda$), the design formula is clearly an excellent approximation for practical purposes. For longer fuses ($l > 4\lambda$), the parabolic TF should not be used since the curves of θ/x no longer resemble parabolas, but resemble curves A and B in figure 5.2. This is no restriction, however, since for all values of l greater than 4λ it is evident that for practical purposes θ_c is substantially equal to θ_∞.

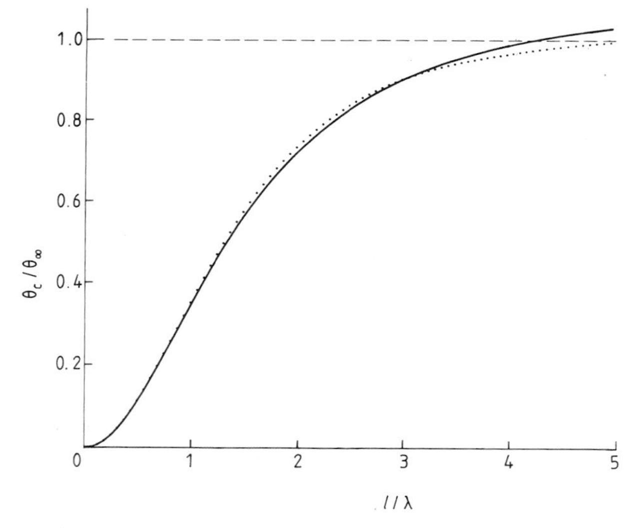

Figure 5.3 Approximate (full curve) and exact (dotted curve) solutions for the temperature θ_c at the centre of the short fuse. Here θ_∞ is the temperature at the centre of a very long fuse and l/λ is a dimensionless measure of the length (see formulae (5.9) and (5.10)).

It will be noticed from the figure that when l is greater than about 4λ the design formula (5.9) gives the physically absurd result that θ_c is greater than θ_∞. It is reassuring to note that this physical indication of the failure of the parabolic hypothesis occurs at about $l = 4\lambda$, exactly the criterion given in §4.4

to mark the change to the long-range problem, for which the exponential hypothesis should be used.

5.3 Collocation of parabolic trial functions

It will be remembered that the aim of collocation is to make two functions agree at as many points as possible, these points being evenly spaced throughout the domain. The parabolic TF is always chosen so that it agrees with the data at the symmetrical boundaries, and at first sight it might seem best to balance the equation at a point halfway between these. Figure 5.4(a) shows that collocation midway between the boundaries divides the x domain into two equal halves, with the collocation point at the extremum.

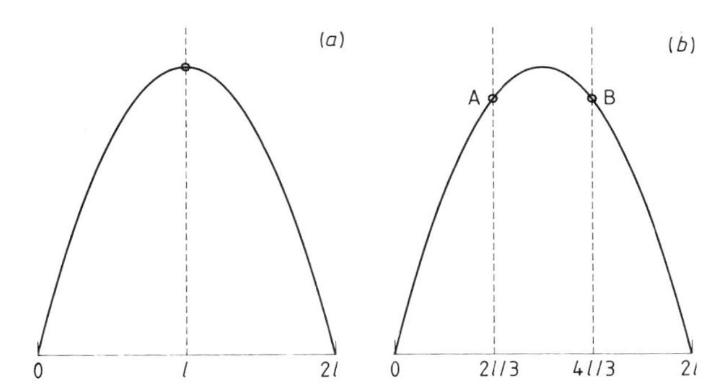

Figure 5.4 Collocation points for parabolic TFs: (a) Collocation midway between the boundaries. (b) Collocation one-third of the way across from the extremum to a boundary. The points A and B are equivalent by symmetry.

The reason why this is not the optimum collocation point can be seen by considering figure 5.4(b), which shows a collocation point A one-third of the way from the extremum towards the left-hand boundary. Because of the mirror symmetry of the solution curve, the equation will then also be balanced at the point B one-third of the way between the extremum and the right-hand boundary. Thus a single collocation according to the rule: 'collocate one-third of the way between extremum and boundary' has the effect of two collocations, at A and B, equally spaced from the boundaries and from each other. The symmetry ensures that collocation at either point gives the same answer. Intuitively this double collocation would be expected to give greater accuracy than the single collocation of figure 5.4(a), and this is confirmed by testing the two alternatives for a large number of equations.

It is possible to argue in favour of more refined rules for choosing the collocation point. Symmetry arguments like the one given above may give different results for two- and three-dimensional equations. Since problems in more than one dimension will not be encountered until PDEs are studied, the appropriate modifications to the rule will be given in Chapter 9 (§9.3). It is also possible for the qualitative sketch to resemble a lopsided parabola, in which case the symmetry argument breaks down.

For most one-dimensional problems in space or time, the 'one-third of the way across' rule gives good results for parabolic TFs, and it can be assumed that the design formulae obtained are numerically useful without further checks. Any breakdown of the parabolic hypothesis is almost always clearly signalled by the physics of the problem, as in the case of the fuse for $l > 4\lambda$.

5.4 The short cooling fin

The cooling fin and its differential equation were introduced in §4.2, where the exponential TF was used to find a design formula for the length constant of a long fin. For $l > 4\lambda$ the temperature $\theta(l)$ at the far boundary has already fallen so close to zero that no further significant loss of heat can occur by lengthening it. Like a fuse, however, a short fin is easier and cheaper to make and takes up less space, but it will evidently dissipate less heat than a long one and the temperature at the far boundary at $x = l$ will not be zero. In this example we shall use a parabolic TF to find a design formula for $\theta(l)$ for a fin of length $l < 4\lambda$.

The governing equation will be taken (see §4.2) as

$$\frac{d^2\theta}{dx^2} - c\theta = 0 \tag{5.11}$$

where

$$c = 2h/Kb. \tag{5.12}$$

Although the convection losses are strictly proportional to θ^n $(1 < n < \frac{5}{4})$, it has already been shown (§4.2.3) that the effect on the temperature distribution of powers of n greater than 1 is slight. The simple, linear assumption that $n = 1$ will therefore be used here for purposes of illustration.

The near boundary condition, as before, is

$$\theta(0) = \theta_0.$$

The far boundary condition depends on the heat loss from the end face. In the simplest model, this is assumed to be zero, so that the far boundary condition is

$$\theta'(l) = 0.$$

In practice, for fins of typical shape, the heat loss from the narrow end face is relatively so small that this is a good approximation.

The qualitative sketch for the short fin is shown in figure 5.5(a). This curve no longer shows the plateau characteristic of the long-range problem, but is

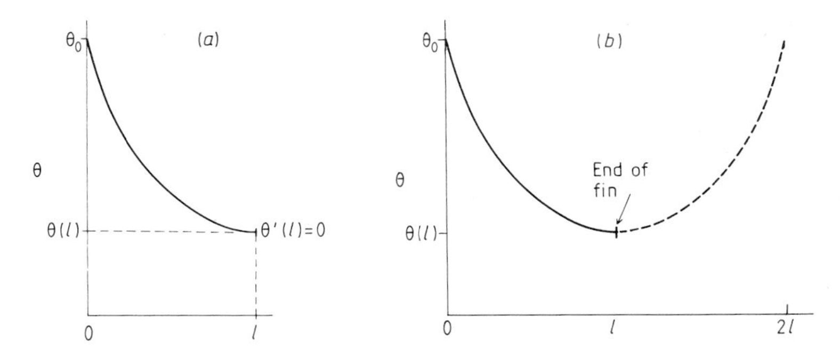

Figure 5.5 (*a*) Qualitative sketch for a short cooling fin. (*b*) Redrawn sketch for the short cooling fin. The fin extends from $x=0$ to $x=l$. The mirror image completes a symmetrical parabola, corresponding to the reference curve in figure 5.1(*b*) (lower).

more closely approximated by half a parabola. In figure 5.5(*b*) the parabola has been completed using mirror symmetry. The figure shows that the boundary condition $\theta'(l)=0$ can be replaced by the equivalent $\theta(2l)=\theta_0$. The qualitative sketch then corresponds with figure 5.1(*b*), with θ replacing u. The corresponding formula (5.1b) then gives the TF

$$\theta^* = \theta_0 + [\theta(l) - \theta_0]\left[1 - \left(\frac{x-l}{l}\right)^2\right]. \tag{5.13}$$

It may be checked that this does indeed satisfy the boundary conditions at $x=0$ and $x=l$. The TF contains one unknown constant $\theta(l)$ to be determined from the equation.

5.4.1 Solution

Substitute the TF into equation (5.11) to obtain the residual equation:

$$\frac{2}{l^2}[\theta(l) - \theta_0] + c\theta_0 + c[\theta(l) - \theta_0]\left[1 - \left(\frac{x-l}{l}\right)^2\right] = \mathscr{R}. \tag{5.14}$$

Collocation at $x=\frac{2}{3}l$ (see figure 5.1(*b*)) gives

$$\frac{2}{l^2}[\theta(l) - \theta_0] + c\{\theta_0 + \tfrac{8}{9}[\theta(l) - \theta_0]\} = 0.$$

This simplifies to

$$\frac{\theta(l)}{\theta_0} = \frac{1 - \frac{1}{18}cl^2}{1 + \frac{4}{9}cl^2}.$$

According to formula (4.7) with $n=1$,

$$c = 1/\lambda^2.$$

Therefore, we have the final design formula for the ratio of the temperature at the far end of the fin to that at the base:

$$\frac{\theta(l)}{\theta_0} = \frac{1-\frac{1}{18}(l/\lambda)^2}{1+\frac{4}{9}(l/\lambda)^2}. \tag{5.15}$$

5.4.2 *Discussion*

Figure 5.6 shows a plot of the design formula (5.15). For fins of length $l=2\lambda$ the temperature $\theta(l)$ at the end is still some 30% of that at the base of the fin, θ_0; if l is increased to 3λ, $\theta(l)$ has fallen to about 10% of θ_0. The precise way in which the heat dissipated depends on $\theta(l)$ needs a further calculation, but intuitively an engineer might feel that figure 5.6 suggests that there would be a significant loss of efficiency if the fin were cut off at $l=2\lambda$, whereas around $l=3\lambda$ there would be little to be gained by further lengthening of the fin.

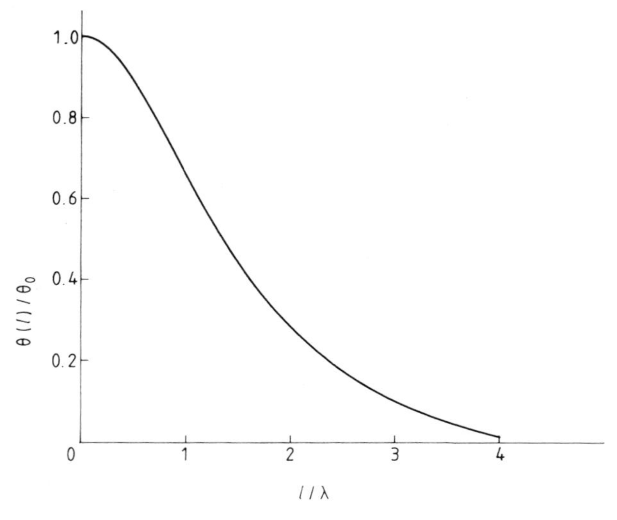

Figure 5.6 Plot of the design formula (5.15) for the ratio of the excess temperature at the end of a short cooling fin to that at the base.

5.5 The period of a large-angle pendulum

In the preceding examples, parabolic TFs have been used to solve ODE 2 with boundary conditions specified; the goal has been to find the height (or depth) of the solution curve. These examples arose as short-range problems, in which solutions of essentially an exponential nature are unable to develop their characteristic plateau because it is cut off by the far boundary.

Parabolic TFs will now be used to study inherently oscillatory equations, whose long-range solutions are not exponentials with plateaux, but oscillations with alternating maxima and minima. The most important goal of an oscillatory equation is a design formula for the period T. The application of the parabolic TF to this end will be illustrated by considering the so-called 'simple' pendulum, swinging through large angles (of the order of half a radian or more).

The governing equation for the angle θ at time t is

$$\frac{d^2\theta}{dt^2} + \left(\frac{g}{l}\right)\sin\theta = 0.$$

$$\left[\begin{matrix}\text{angular}\\\text{acceleration}\end{matrix} + \begin{matrix}\text{restoring torque}\\\div\text{moment of inertia}\end{matrix} = 0\right] \tag{5.16}$$

where g is the acceleration due to gravity and l is the length from the pivot to the centre of mass of the bob. The initial conditions are

$$\theta(0) = \theta_0 \qquad \theta'(0) = 0.$$

These state that the pendulum is released from an initial angle θ_0, the initial angular velocity being zero. The goal is to find a design formula for the period T as a function of the amplitude θ_0.

The pendulum is an object of common experience, and the qualitative sketch (figure 5.7) of the oscillations of θ must show the sequence of alternating maxima and minima, corresponding to the bob's coming to rest ($\theta' = 0$) alternately on the left and on the right.

Further information about the shape of the curve is obtained from the equation. If θ is small, the linear approximation $\sin\theta \approx \theta$ reduces equation (5.16) to the linear equation

$$\frac{d^2\theta}{dt^2} + \left(\frac{g}{l}\right)\theta = 0$$

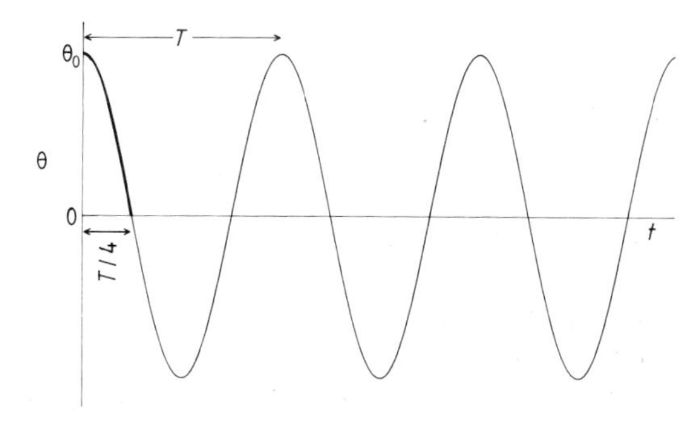

Figure 5.7 Qualitative sketch for the large-amplitude pendulum. The first quarter oscillation may be approximated by a parabola.

which has the solution (§2.4)

$$\theta=\theta_0 \cos\left[\frac{1}{2\pi}\left(\frac{g}{l}\right)^{1/2} t\right]$$

for the stated initial conditions. The maximum value of θ is set by the initial value θ_0, and it may be inferred that for modest values of θ_0 the solution of equation (5.16) will not differ markedly from the cosine solution of the linear equation; figure 5.7 is the qualitative sketch drawn on this basis. If the first quarter period only is considered, however, the curve looks like a half-parabola with its extremum at $t=0$. This short-range solution, emphasised by the heavy full curve in figure 5.7, is redrawn in figure 5.8 to show its visual relation to figure 5.1(a), and hence the possibility of using the parabolic TF (5.1a). Writing θ for u, t for x and the quarter-period $\frac{1}{4}T$ for l, the TF given by the corresponding formula (5.1a) is

$$\theta^* = \theta_0\left[1-\left(\frac{4t}{T}\right)^2\right].$$

θ_0 is given by the first initial condition, so the only unknown constant in the TF is T, the period for a complete oscillation.

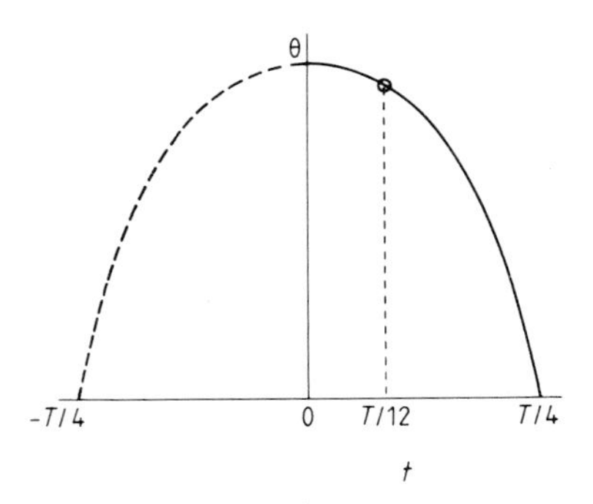

Figure 5.8 Parabolic sketch for the large-amplitude pendulum, completed to facilitate comparison with the reference curves in figure 5.1. The collocation point is also shown (small open circle).

5.5.1 Solution

Substituting the TF into the governing equation (5.16), we obtain the residual equation:

$$\frac{-32\theta_0}{T^2}+\left(\frac{g}{l}\right)\sin\left\{\theta_0\left[1-\left(\frac{4t}{T}\right)^2\right]\right\}=\mathscr{R}.$$

Collocating one-third of the way across (figure 5.8 shows this to be at $t=T/12$), we find the result is

$$\frac{-32\theta_0}{T^2}+\left(\frac{g}{l}\right)\sin\left(\tfrac{8}{9}\theta_0\right)=0.$$

Solving for the period then gives

$$T=4\left(\frac{2\theta_0}{\sin\left(\frac{8}{9}\theta_0\right)}\right)^{1/2}\left(\frac{l}{g}\right)^{1/2}$$

which can be written as

$$\frac{T}{T_0}=\frac{2}{\pi}\left(\frac{2\theta_0}{\sin\left(\frac{8}{9}\theta_0\right)}\right)^{1/2} \tag{5.17}$$

where $T_0=2\pi(l/g)^{1/2}$ is the well known exact solution for θ_0 very small.

5.5.2 Discussion

Figure 5.9 shows the design formula (5.17) plotted for values of θ_0 between 0° and just less than 180°, the maximum practical value. The exact result, shown for comparison, can only be found using elliptic integrals.

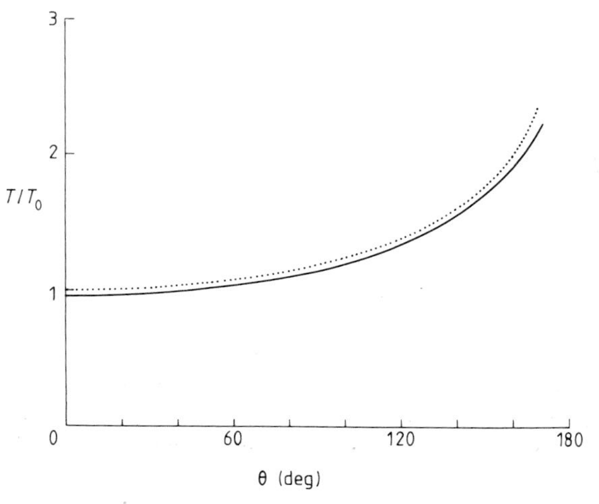

Figure 5.9 Period of a large-angle pendulum calculated from the design formula (5.17) (full curve), with the exact result for comparison (dotted curve).

It will be seen that the design formula gives a useful idea of the way in which T depends upon the initial angle θ_0 up to about $\theta_0=170°$, when the bob is being released from a position nearly vertically above the pivot. However, failure at the largest angles can be anticipated from physical considerations.

When the pendulum bob is released from a position exactly vertically above the pivot ($\theta_0 = 180°$), it will not move, since there is no resultant horizontal force either way. Physical reasoning therefore tells us that $T = \infty$ when $\theta_0 = 180°$. On the other hand, the design formula (5.17) predicts that T does not become infinite until sin $(\frac{8}{9}\theta_0) = 0$, that is $\theta_0 = \frac{9}{8} \times 180°$, or about 202°: this is an angle well beyond the physically meaningful limit of 180°. The breakdown is signalled for large values of θ_0 by this physical absurdity in a way very similar to the breakdown of the parabolic hypothesis for the long fuse (see end of §5.2). The cause of the breakdown is again that the shape of the solution curve is too far removed from a parabola to allow a parabolic TF to be a useful hypothesis.

Problems

5.1

In §5.2 the governing equation for a short fuse was based on a linear heat loss term $a\theta$. If the heat loss is assumed to be proportional to θ^n, equation (5.2) becomes

$$-\frac{d^2\theta}{dx^2} + a\theta^n = b. \qquad (5.18)$$

(a) Repeat the work of §5.2, using the same TF, to find the collocation equation (corresponding to equation (5.7)) for the more general equation (5.18).

(b) For fuses, the design formula that is required is the length $2l$ to attain a given value of θ_c at a given current. Solve the equation obtained in (a) for l in terms of θ_c, a and b.

(c) Noting that the term involving the constant a corresponds to convective loss, which is relatively small when the fuse is short, use an 'a small' approximation (§2.8) to simplify the design formula for l to two additive terms on the RHS. (In this, the first term corresponds to the effect of conduction to the supports and the second to a correction for convective effects.)

5.2

The governing equation for steady laminar (nonturbulent) flow of a liquid through a circular pipe of radius R is (see §10.2 for details)

$$\frac{1}{r}\frac{d}{dr}\left(r\frac{dv}{dr}\right) + \frac{p}{\eta L} = 0 \qquad (5.19)$$

where v is the velocity along the pipe at a radial distance r from the axis.

At the walls $v = 0$, so intuitively the possible qualitative sketches for the velocity profile v/r would be expected to be like figure 5.2, with a plateau in the central portion for large-diameter pipes. On the other hand, the parabolic TF would be appropriate for small-diameter pipes, in which the influence of the walls extends to the centre.

(a) Show from the equation that a plateau cannot in fact develop. [End of §2.3] There is therefore no point in trying the exponential as a TF.

(b) Write down a parabolic TF that satisfies the boundary condition $v(R)=0$ and contains the unknown parameter v_c, the velocity at the centre ($r=0$) of the pipe.

(c) Substitute, and after noting that the residual can be made zero for all r, find a design formula for v_c.

(The TF turns out to be exact, since it balances the equation for all r. The solution can be found quite easily by two successive integrations of equation (5.19); the interesting point made by thinking about the example from the point of view of the TF method is that the intuitive idea of a plateau for large pipes is not fulfilled, the effect of the walls being apparently felt at the centre however large the diameter. It is reassuring that the impossibility of a plateau was detected at an early stage, when it was seen that there was no solution for v_∞.)

5.3

The equation for a ball-bearing oscillating in a glass tube cited in §1.2 is

$$\frac{d^2u}{dt^2}+cu^3=0.$$

The goal is a design formula for the period and hence the frequency. Assume that the initial conditions are $u(0)=u_0$ and $u'(0)=0$.

(a) Write down a TF containing the period T as the unknown parameter. [Figures 5.7 and 5.8; §5.5]

(b) Substitute and collocate to find a design formula for T.

(c) Write down the design formula for the oscillation frequency f.

(The almost identical result (1.10) quoted in §1.2 was obtained by an alternative method using a sinusoidal TF (see problem 6.2). The exact result obtained using elliptic integrals has the same form but a numerical constant 0.135.)

5.4

A problem of some importance in plasma physics concerns an electron beam injected into a plasma tube where the magnetic field is cylindrical and increases towards the axis in inverse proportion to the radius. The beam is injected parallel to the axis, but the magnetic field bends the path towards the axis. The beam first crosses the axis at a distance L from the point of entry, as shown in figure 5.10.

The governing equation for the path $y(x)$ of the electrons is

$$\frac{d^2y}{dx^2}+\frac{c}{y}=0$$

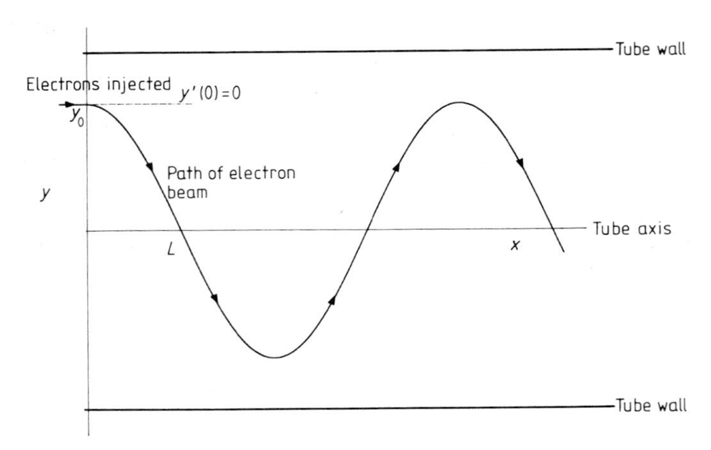

Figure 5.10 Qualitative sketch of the path of an electron beam injected into a long cylindrical plasma tube with nonuniform axial magnetic field.

where c is a constant for a given magnetic field. The initial conditions are $y(0)=y_0$ and $y'(0)=0$.

(a) It is not intuitively obvious what kind of path the beam will follow. Show that the equation describes an oscillation in space. [§2.6] The beam therefore crosses and recrosses the x-axis as it passes through the device (figure 5.10).

(b) The goal is to find L, the value of x where the beam first crosses the x-axis. Write down a parabolic TF with L as the unknown parameter.

(c) Find a design formula for L.

(The exact result obtained in terms of the error function is the same except the numerical factor is 1.25.)

5.5

The equation for the potential V in a planar thermionic diode is

$$\frac{d^2V}{dx^2}=\frac{kj}{V^{1/2}}$$

where x is the distance from the cathode, j is the current per unit area and k is a known constant. The auxiliary conditions are $V(0)=0$ and $V'(0)=0$, and the potential applied to the anode situated at $x=L$ is V_a. The goal is to find a design formula connecting j and V_a.

(a) Make a qualitative sketch of V against x, and confirm that the parabolic TF

$$V^*=V_a\frac{x^2}{L^2}$$

matches the sketch.

(b) Substitute, and collocate at $x = L/3$, to obtain the design formula

$$j = \frac{2}{3} \frac{V_a^{3/2}}{kL^2}.$$

(The exact result has the same functionality, but the numerical factor is $\frac{4}{9}$ instead of $\frac{2}{3}$. This numerical error of 50% is unusually large, and is a consequence of the peculiar term involving $V^{-1/2}$. Quite generally, for equations with terms having negative powers or logarithms of the variables, the choice of TF requires special care. [See §§9.6 and 11.3])

Chapter 6

The QSTF method for unforced oscillations

6.1 Realistic oscillator equations

The simple linear equation

$$\underset{\text{[acceleration]}}{\frac{d^2u}{dt^2}} + \underset{\text{[restoring force]}}{\omega_0^2 u} = 0 \qquad (6.1)$$

is often used as an approximate physical equation to describe an oscillator. The frequency of oscillation is then given by (§2.4)

$$f = \omega_0/2\pi.$$

Since ω_0 is a constant of the physical system, the frequency is therefore predicted to be a constant independent of the amplitude. This prediction is never precisely fulfilled experimentally, for in all real oscillators the frequency is observed to depend on the amplitude of oscillation. The simple linear model embodied in equation (6.1) is unable to explain this phenomenon; it is also useless for studying how oscillations can build up from near zero to a steady-state amplitude when a free-running oscillator is switched on.

To make the mathematical model more realistic, a damping term (§2.5) must be included; moreover, it must be recognised that in practical systems both the damping coefficient and the restoring term coefficient may depend on u, so the equations are nonlinear. A more realistic equation, which can describe many real oscillators, is therefore

$$\underset{\text{[acceleration]}}{\frac{d^2u}{dt^2}} + \underset{\text{[damping]}}{p(u)\frac{du}{dt}} + \underset{\text{[restoring]}}{q(u)u} = 0. \qquad (6.2)$$

This equation is similar to the linear damped oscillator equation (2.21) except that the coefficients may now be functions of u. For the oscillations to continue indefinitely at constant amplitude, the damping coefficient $p(u)$ must not be positive throughout the cycle, causing the oscillations to decay; nor must it be negative throughout the cycle, causing them to build up.

It follows that realistic oscillator equations can be usefully solved with sinusoidal TFS in three different circumstances:

(a) $p(u)=0$ (exceptional, but see §6.5).

(b) $p(u)$ is so small that the amplitude is substantially constant over many cycles. Equation (6.2) can then be used with $p(u)$ taken as zero.

(c) $p(u)$ varies so that it is positive for one part of the cycle and negative for the other. It may be possible for the net damping effect, averaged over a complete cycle, to be zero, so that the amplitude remains constant indefinitely. Practical examples of all these possibilities will be studied in this chapter using sinusoidal TFS.

It is convenient for the remaining discussion to simplify the notation of equation (6.2) by writing $Q(u)=q(u)u$, so that the equation can be written in the form

$$\frac{d^2u}{dt^2}+\underbrace{p(u)\frac{du}{dt}}_{\text{[damping]}}+\underbrace{Q(u)}_{\text{[restoring]}}=0. \qquad (6.3)$$

The QSTF method is particularly advantageous in understanding models of unforced oscillators, since equations of the form (6.3) are generally impossible to solve exactly; they are usually treated either by oversimplification to the unsatisfactory linear form (6.1), or by approximate methods too complicated for the nonspecialist to understand. The QSTF method allows realistic models of oscillators to be solved approximately in a direct and intelligible way. Furthermore, it can provide useful design formulae for the relation between frequency and amplitude, which are the two properties of greatest physical importance in characterising oscillations.

6.2 Cosine trial functions

Although the waveforms of real oscillators may differ significantly from pure sinusoids, simple sinusoidal TFS are usually sufficiently good approximations to give useful design formulae for the fundamental frequency and amplitude. Once it is established that the equation describes an unforced oscillation, a sinusoid is always chosen for the qualitative sketch, and the choice of TF is limited to two possibilities:

the *cosine TF* (figure 6.1(*a*))

$$u^*=u_1 \cos \omega t \qquad (6.4)$$

and the *shifted cosine TF* (figure 6.1(*b*))

$$u^*=u_0+u_1 \cos \omega t. \qquad (6.5)$$

The constant u_0 is familiar in electronics as the DC component resulting from

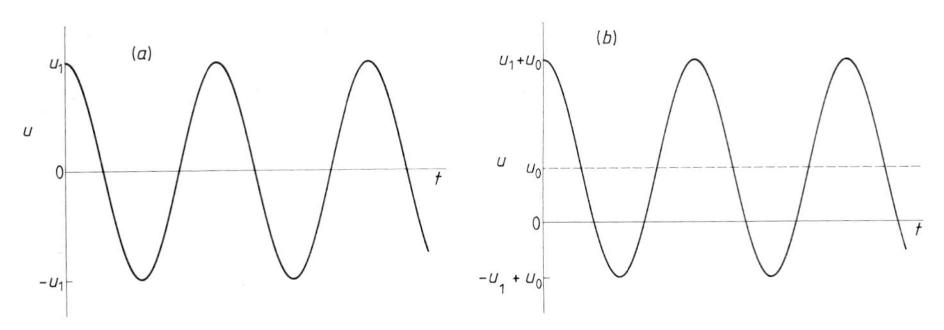

Figure 6.1 Qualitative sketches for steady-state oscillations; (*a*) cosine; (*b*) shifted cosine.

the rectifying action associated with square-law devices. Alternatively, it can be regarded as the time-averaged value of u, since the average of the cosine part of the TF taken over a cycle is zero.

The choice between the two TFs is made by inspection of the restoring term in equation (6.3). If the restoring term $Q(u)$ is an odd function of u (that is, it only contains odd powers of u) then the simpler function (6.4) is used. If the restoring term contains an even power, as for instance

$$Q(u)=\alpha u+\beta u^2 \tag{6.6}$$

then the shifted cosine function (6.5) must be used.

The reason for this rule is that if $Q(u)$ is an odd function, the average restoring term $\overline{Q(u)}$, taken over one complete cycle, is zero. The average displacement u_0 must therefore also be zero. If $Q(u)$ contains any even powers of u, its average value is no longer zero, and the average displacement u_0 is therefore also nonzero (see problem 6.1).

A special case of an even power in $Q(u)$ arises when there is a constant c instead of zero on the RHS of equation (6.3). A constant c may be regarded as cu^0 (since $u^0=1$), and therefore as contributing an even-power term. It follows that if an oscillatory equation contains a constant forcing term, the shifted cosine TF must be used (see problems 6.4 and 6.5).

Since the qualitative sketch for unforced oscillations is always a sinusoid, we shall not describe the qualitative sketch in the examples that follow. Once the TF is chosen, the corresponding qualitative sketch is also automatically chosen as one of the two curves in figure 6.1. One other difference in procedure is that for sinusoidal TFs the residual is best minimised not by collocation but by the method of harmonic balance, which will be explained in the examples and discussed in §6.4.

6.3 Nonlinear restoring term: the small-angle pendulum

The motion of a pendulum with finite angle of swing has already been discussed in §5.5 using a parabolic TF. Here we shall treat the problem by using a sinusoidal TF; the objective will be limited to finding how the period of oscillation depends on amplitude when the angle of swing is small, that is not more than 0.5 rad. The governing equation (assuming damping is so small as to be negligible) is

$$\frac{d^2\theta}{dt^2}+\left(\frac{g}{l}\right)\sin\theta=0.$$

A dimensionless timescale is conveniently introduced (see appendix 3) by writing

$$x=(g/l)^{1/2}t. \tag{6.7}$$

The equation is then

$$\frac{d^2\theta}{dx^2}+\sin\theta=0.$$

For values of θ less than 0.5 rad, this may be approximated by

$$\frac{d^2\theta}{dx^2}+\theta-\tfrac{1}{6}\theta^3=0 \tag{6.8}$$

and so the restoring term is $Q(\theta)=\theta-\frac{1}{6}\theta^3$.

6.3.1 Solution

(i) Choice of TF

Since there is no damping term in the equation being used as the mathematical model, the oscillations will continue indefinitely at a constant amplitude. Equation (6.8) therefore describes a free oscillator, and a cosine TF can be used. Since

$$Q(\theta)=\theta-\tfrac{1}{6}\theta^3$$

is an odd function of θ, the rule given in §6.2 shows that the appropriate TF for steady-state oscillations is the unshifted cosine

$$\theta^*=\theta_0\cos\omega x. \tag{6.9}$$

Here θ_0 is the initial amplitude when the pendulum is released; ω is the single unknown parameter, and the goal is to find a design formula relating ω to θ_0.

(ii) Substitution to form the residual equation

The TF (6.9) substituted into equation (6.8) gives

$$-\omega^2\theta_0\cos\omega x+\theta_0\cos\omega x-\tfrac{1}{6}\theta_0^3\cos^3\omega x=\mathcal{R}.$$

(iii) Harmonic balancing

(A more detailed discussion of this step will be given in the following section.)

Any powers of sines or cosines must first be replaced by the equivalent functions of multiple angles. Here we write†

$$\cos^3 \omega x = \tfrac{1}{4}(3 \cos \omega x + \cos 3\omega x).$$

Substitution for $\cos^3 \omega x$ and collection of terms simplifies the residual equation to

$$(-\omega^2\theta_0 + \theta_0 - \tfrac{1}{8}\theta_0^3) \cos \omega x - \tfrac{1}{24}\theta_0^3 \cos 3\omega x = \mathscr{R}. \tag{6.10}$$

It is now assumed that the second term, involving $\cos 3\omega x$, is relatively small and may be neglected, so that the residual can be approximately minimised by making the first term zero. This is achieved by making the coefficient of $\cos \omega x$ zero; that is

$$-\omega^2\theta_0 + \theta_0 - \tfrac{1}{8}\theta_0^3 = 0.$$

Hence

$$\omega = (1 - \tfrac{1}{8}\theta_0^2)^{1/2}.$$

Since $\theta_0 < \frac{1}{2}$, the correction term $\theta_0^2/8$ is much less than unity. Therefore the design formula for ω is

$$\omega = 1 - \tfrac{1}{16}\theta_0^2 \tag{6.11}$$

by the binomial theorem for 'θ_0 small'.

Using $T = 2\pi/\omega$ and relation (6.7), we obtain as the design formula for the period

$$T = 2\pi \left(\frac{l}{g}\right)^{1/2} (1 - \tfrac{1}{16}\theta_0^2)^{-1}$$

or

$$T = 2\pi \left(\frac{l}{g}\right)^{1/2} (1 + \tfrac{1}{16}\theta_0^2) \tag{6.12}$$

again making use of the binomial theorem to obtain a 'θ_0 small' approximation. This is identical to the result obtained if elliptic integrals are used to solve equation (6.8) exactly.

6.3.2 *Discussion*

As θ_0 tends to zero, formula (6.12) predicts that

$$T \rightarrow 2\pi \left(\frac{l}{g}\right)^{1/2}$$

† A collection of the most useful conversion formulae for this purpose is contained in appendix 6.

which is the exact result obtained using a linear model. However, the design formula shows that, as the amplitude θ_0 increases, the period lengthens, and that even when θ_0 is as small as 0.1 rad (5.7°), T has increased by nearly 0.1%. In time-keeping applications this is a considerable change, amounting to about 1 min per day. This explains why pendulum clocks must be designed so that the amplitude of oscillation is kept small and also as constant as possible.

6.4 Residual minimisation by harmonic balance

In the TF method applied to nonoscillatory equations, minimisation of the residual was achieved by collocation at a single representative point on the solution curve. For cosine TFs, harmonic balance is usually a simpler method, and gives results that are as good as or better than those obtained by collocation.

The *method of harmonic balance* is to equate to zero the coefficients of as many harmonic terms as there are adjustable parameters in the TF. In the simplest cases, there will be a single adjustable parameter, enabling only the single fundamental term to be balanced, as in §6.3. In more complicated problems, two parameters may have to be adjusted. This may involve either balancing the zero-frequency (constant) term and the fundamental, or balancing two terms at the fundamental frequency, one containing $\sin \omega x$ and the other containing $\cos \omega x$. There is no limit in principle to the number of harmonics that may be balanced, but solutions using TFs containing more than two or three adjustable parameters tend to be clumsy and too complicated for easy understanding.

Balancing the lower harmonics by setting their coefficients to zero gives design formulae that are sometimes exact, and generally more accurate than collocation. The success of the method can most simply be explained by regarding the LHS of the residual equation as the start of a Fourier series. If this series converges sufficiently rapidly for the higher terms to be relatively small, they may be neglected and the equation balanced approximately for all x by balancing only the lowest terms. To find n unknowns in the TF, n equations are needed, and so the first n coefficients of the terms in the series must be equated to zero.

To summarise the steps of the harmonic balance procedure for cosine TFs:

(a) Replace powers and products of sines and cosines by multiple angles.
(b) Count the number of unknown parameters in the TF.
(c) Balance a sufficient number of the harmonics, starting at the lowest frequency, to determine these parameters.

A method of checking the results of harmonic balance will be described in §6.7.

For some equations, step (a) may not be practicable. For example, if a negative or fractional power of a sine or cosine appears, it is not possible to express it simply in terms of multiple angles. In general, the method of harmonic balance is useful when the restoring term contains positive integral powers of u; this covers most cases of practical interest (see also problem 6.3).

6.5 Asymmetric restoring term: atomic vibrations

Atoms are bound together in a solid by a mutual potential energy, which can be expressed in terms of the distance R between the atoms by means of a function such as

$$W=\frac{A}{R^2}+\frac{B}{R^6}.$$

The equilibrium distance R_0 is such that W is a minimum, W_0. For small displacements u from R_0, the potential energy may be written as

$$W-W_0=au^2-bu^3.$$

Since $F=-\mathrm{d}W/\mathrm{d}u$, the restoring force acting on an atom is given by

$$F=-2au+3bu^2.$$

The equation of motion for atoms of mass m is therefore

$$m\frac{\mathrm{d}^2u}{\mathrm{d}t^2}+2au-3bu^2=0.$$

This may be simplified to

$$\underset{\text{[acceleration]}}{\frac{\mathrm{d}^2u}{\mathrm{d}t^2}} \quad \underset{\text{[restoring]}}{+\,\alpha u-\beta u^2}=0 \tag{6.13}$$

where $\alpha=2a/m$ and $\beta=3b/m$. In the notation of §2.6, $qu=\alpha u-\beta u^2$. Provided that $\beta u<\alpha$, a condition always satisfied in stable solids, $q=\alpha-\beta u$ is always positive and the atoms must therefore oscillate about their mean positions. A prime goal in solving this mathematical model is to find a design formula for the average atomic separation as a function of the amplitude of the oscillations.

6.5.1 Solution

Since the restoring term $Q(u)=\alpha u-\beta u^2$ contains an even power, the TF according to the rule in §6.2 is

$$u^*=u_0+u_1\cos\omega t.$$

There are two unknown parameters, u_0 and ω; u_1 is not an unknown parameter but is known independently from statistical mechanics (see the discussion below). Parameter u_0 is the average shift in the interatomic spacing as a consequence of the oscillations, and the goal is to find a design formula for u_0 in terms of u_1.

Substituting u^* into equation (6.13) gives

$$-\omega^2 u_1 \cos \omega t + \alpha u_0 + \alpha u_1 \cos \omega t - \beta u_0^2 - 2\beta u_0 u_1 \cos \omega t - \beta u_1^2 \cos^2 \omega t = \mathscr{R}.$$

Since (from appendix 6)

$$\cos^2 \omega t = \tfrac{1}{2} + \tfrac{1}{2} \cos 2\omega t$$

the residual equation, arranged in ascending order of frequency, is

$$\alpha u_0 - \beta u_0^2 - \tfrac{1}{2}\beta u_1^2 + (-\omega^2 + \alpha - 2\beta u_0) u_1 \cos \omega t + (-\tfrac{1}{2}\beta u_1^2) \cos 2\omega t = \mathscr{R}. \tag{6.14}$$

For two unknown parameters, two harmonic terms must be balanced, starting at the lowest frequency, so that

$$\alpha u_0 - \beta u_0^2 - \tfrac{1}{2}\beta u_1^2 = 0 \qquad \text{(constant)} \tag{6.15}$$

and

$$(-\omega^2 + \alpha - 2\beta u_0) = 0 \qquad \text{(fundamental).} \tag{6.16}$$

Equation (6.15) is a quadratic but can be more usefully solved by noting that in practice $\beta u_0/\alpha \ll 1$ so that the second term $-\beta u_0^2$ is negligible. To a sufficient approximation, therefore,

$$u_0 = \tfrac{1}{2}(\beta/\alpha) u_1^2. \tag{6.17}$$

Parameter u_0 is the shift in the average atomic spacing caused by the oscillations of amplitude u_1, and formula (6.17) is the design formula that is the goal of the problem. It is not necessary for the purpose of understanding thermal expansion to solve equation (6.16) to find ω.

6.5.2 *Thermal expansion*

According to formula (6.17), the change in the average distance between atoms obeying the assumed law is proportional to u_1^2. The kinetic energy of the atoms is also proportional to u_1^2, and according to statistical mechanics the mean kinetic energy is proportional to absolute temperature. The model therefore correctly predicts that the mean distance between atoms increases approximately linearly with temperature; this is the phenomenon of linear thermal expansion. The experimentally observed small departures from this linear relationship are caused by higher-order nonlinear terms in the force law.

6.6 Nonlinear damping: maintained oscillations of a van der Pol oscillator

Most electronic oscillators have a common purpose: the production of stable oscillations at a controlled frequency and amplitude. In order to make up for the inevitable energy dissipation and energy drawn from the oscillator by an external load, it is necessary to supply energy from a DC source. This process of balancing energy loss from an oscillator by supplying steady (DC) excitation is called *maintained oscillation.* It is quite different from *forced oscillation,* discussed in the next chapter, in which one oscillator is driven by another. The most famous mathematical model of maintained electronic oscillators is due to van der Pol. Although originally derived for a particular type of thermionic valve oscillator, the van der Pol equation applies approximately to a wide range of electronic oscillators containing a tuned circuit, and also to certain mechanical oscillators.

The *van der Pol equation* can be written

$$\frac{d^2u}{dx^2} \underbrace{-\varepsilon(1-au^2)\frac{du}{dx}}_{\text{[damping]}} + \underbrace{\omega_0^2 u}_{\text{[restoring]}} = 0. \tag{6.18}$$

The damping term consists of two parts: a linear, negative term $-\varepsilon\, du/dx$, which tends to increase the amplitude of oscillation; and a nonlinear, positive term $\varepsilon au^2\, du/dx$, which tends to decrease the amplitude. It is these two opposing terms in nonlinear damping that allow an explanation of the building of oscillations to a final stable level. When the oscillator is first switched on, the amplitude is very small, being due to the residual noise in the system; u is therefore initially close to zero, and $au^2 \ll 1$. The damping is accordingly negative at first, so the oscillations grow in amplitude. Eventually the effect of the positive damping term is sufficient to balance the effect of the negative term, and the amplitude of oscillation reaches a steady value. It is clear from the qualitative discussion that nonlinear damping is essential for the production of stable, maintained oscillations. Constant, linear damping inevitably leads either to the eventual disappearance or unlimited growth of oscillations, according to the sign of the damping term.

The primary goals in solving the van der Pol equation are to find the frequency and amplitude of the steady oscillations. The TF method will now be applied to find design formulae for these quantities. It is assumed that $\varepsilon \ll 1$, a condition often satisfied in practice.

6.6.1 *Solution*

The restoring term

$$Q(u) = \omega_0^2 u$$

contains only an odd power of u. The TF for steady-state oscillations is therefore

$$u^* = u_1 \cos \omega x$$

containing two unknown constants u_1 and ω. Substitution of u^* into equation (6.18) gives

$$-\omega^2 u_1 \cos \omega x + \varepsilon(1 - a u_1^2 \cos^2 \omega x)\omega u_1 \sin \omega x + \omega_0^2 u_1 \cos \omega x = \mathscr{R}.$$

The product of $\cos^2 \omega x$ and $\sin \omega x$ is expressed in multiple angles by writing (appendix 6)

$$\cos^2 \omega x \sin \omega x = \tfrac{1}{4} \sin \omega x + \tfrac{1}{4} \sin 3\omega x.$$

The residual equation, arranged in ascending order of frequency, is therefore

$$(\omega_0^2 - \omega^2)u_1 \cos \omega x + \varepsilon \omega u_1 (1 - \tfrac{1}{4} a u_1^2) \sin \omega x - \tfrac{1}{4} \varepsilon \omega a u_1^3 \sin 3\omega x = \mathscr{R}.$$

Since there are two adjustable parameters in the TF, two coefficients in the residual equation must be equated to zero; these are the coefficients of the two lowest-frequency terms, containing $\cos \omega x$ and $\sin \omega x$. Equating the first coefficient to zero gives $\omega = \omega_0$ and therefore

$$f = \omega_0 / 2\pi \tag{6.20}$$

and from the second term $1 - \tfrac{1}{4} a u_1^2 = 0$ or

$$u_1 = 2a^{-1/2}. \tag{6.21}$$

These are the required design formulae for the stable frequency and amplitude.

6.6.2 *Discussion*

Formula (6.20) states that the steady-state frequency of maintained oscillations is not affected by the damping term, being exactly the same as the frequency of the simple harmonic oscillator equation (6.1). A more rigorous but difficult treatment of the problem by perturbation shows that this result is correct to order ε, but that correction terms of order ε^2 do occur. Since ε has been assumed to be small, this correction is very small.

Formula (6.21) (which is identical to that obtained by the perturbation method) states that the stable amplitude is controlled by the parameter a. This parameter is a constant for a given oscillator; for an electronic oscillator, its value can be found from the I–V characteristic of the active device used to maintain oscillation.

6.7 The minimum residual and the residual ratio test*

The numerical usefulness of design formulae obtained as a result of residual minimisation by collocation can be checked by varying the collocation point.

* This section may be omitted at a first reading.

This method cannot be used to check design formulae obtained by harmonic balance. The check proposed here is to calculate a quantity called the residual ratio. After harmonic balance has been achieved, the residual will be a minimum for a simple cosine function; we call its value the *minimum residual* and denote it by $\mathscr{R}_{\min}$. The *residual ratio* ρ is defined as

$$\rho = \frac{\text{amplitude of } \mathscr{R}_{\min}}{\text{amplitude of restoring term}}. \tag{6.22}$$

The smaller the value of ρ, the closer the design formula is to the exact result, and the closer the TF solution is to the exact solution. A simple rule that works well for most practical problems is the residual ratio test:

> 'Design formulae obtained by harmonic balance using oscillatory TFs are likely to be numerically useful if the residual ratio is less than 0.3.'

This rule is based on the criterion (§1.5) that a design formula is numerically useful if the error is less than $\pm 30\%$. It does not imply a strict relationship between the value of ρ and the error in a design formula, but as the following discussion will show there is a strong correlation between them.

Consider the linear harmonic oscillator equation

$$\frac{d^2u}{dx^2} + \omega_0^2 u = 0.$$

If the TF method is applied to this by writing

$$u^* = u_1 \cos \omega x$$

then the residual equation is

$$(-\omega^2 + \omega_0^2)u_1 \cos \omega x = \mathscr{R}.$$

Balancing the fundamental by making $\omega = \omega_0$ reduces the residual to zero for all values of x. Therefore

$$\mathscr{R}_{\min} = 0 \qquad \text{(all } x\text{)}.$$

The vanishing of the residual for all values of x shows that in this case the TF solution is exact.

For nonlinear equations, the minimum residual never vanishes for all values of x. Consider the simple pendulum equation (6.8). The residual equation (6.10) is

$$(-\omega^2 + 1 - \tfrac{1}{8}\theta_0^2)\theta_0 \cos \omega x - \tfrac{1}{24}\theta_0^3 \cos 3\omega x = \mathscr{R}.$$

Balancing the fundamental leaves the minimum residual

$$\mathscr{R}_{\min} = -\tfrac{1}{24}\theta_0^3 \cos 3\omega x. \tag{6.23}$$

The amplitude of the restoring term in equation (6.8) is its peak value, which is approximately θ_0, since $\frac{1}{6}\theta_0^3 \ll \theta_0$. Therefore, from the definition (6.22), the residual ratio is calculated to be

$$\rho = \frac{\frac{1}{24}\theta_0^3}{\theta_0} = \frac{\theta_0^2}{24}.$$

Since θ_0 was originally specified as less than 0.5, ρ in this example is less than about 0.01.

The relationship between the residual ratio ρ and the accuracy of design formulae can be understood qualitatively by the following argument. Consider the residual (6.23): since $\mathscr{R}_{\min}$ is proportional to $\cos 3\omega x$, it might be supposed that it can be cancelled by including a second term $\theta_1 \cos 3\omega x$ in the TF, so that

$$\theta^* = \theta_0 \cos \omega x + \theta_1 \cos 3\omega x.$$

This addition to the TF would produce an extra term proportional to $\theta_1 \cos 3\omega x$ in the residual equation. A value of θ_1 could then be found which balances the third harmonic term, thus reducing $\mathscr{R}_{\min}$. If $\rho \ll 1$, $\mathscr{R}_{\min}$ is already small and so the value of θ_1 required to balance the third harmonic term must be small relative to the fundamental amplitude θ_0. Consequently if $\rho \ll 1$ it follows that $\theta_1 \ll \theta_0$.

Balancing the third harmonic term in the residual equation by this means would still not make $\mathscr{R}_{\min}$ zero, however, because terms containing still higher harmonics of ω would be introduced into the residual equation. By continuing to add terms to θ^* it is possible in principle to balance each higher harmonic to zero. The TF necessary to achieve this would be of the form

$$\theta^* = \theta_0 \cos \omega x + \theta_1 \cos 3\omega x + \theta_2 \cos 5\omega x + \cdots.$$

The argument given above applies as each new harmonic is added and suggests that

$$\theta_0 \gg \theta_1 \gg \theta_2 \ldots \qquad \text{if } \rho \ll 1.$$

In words, if the residual ratio is small, the exact solution is the sum of a series of harmonics, which is so rapidly convergent that balancing only the first one or two terms can provide a useful approximation to the exact solution. These are the circumstances in which one would expect the use of simple cosine TFs to lead to useful results.

Experience confirms that the likely fractional error and the residual ratio are well correlated, and that the fractional errors in the design formulae are usually less than the residual ratio. Thus the residual ratio $\rho < 0.01$ in the case of the small-angle pendulum implies a numerical accuracy of better than 1% in the design formula (6.12) for the period of the pendulum. On the other hand, when ρ exceeds about 0.3, the indication is that the numerical error in the design formula is too great for it to be used without great caution.

Problems

6.1

(a) Confirm that if $Q(u)=u^3$ (an odd function) and $u=\cos \omega t$, then the integral of $Q(u)$ taken over a complete cycle

$$\int_0^T Q(u)\,\mathrm{d}t=\int_0^{t=2\pi/\omega} (\cos \omega t)^3\,\mathrm{d}t$$

is zero. [First convert $(\cos \omega t)^3$ to terms in multiple angles (appendix 6)]

(b) If $Q(u)=u^2$ (an even power of u), confirm that the integral

$$\int_0^T Q(u)\,\mathrm{d}t$$

is not zero.

(The distinction between the integrals of odd and even powers of u is generally true, and leads to the rule of §6.2 for choosing TFS.)

6.2

According to the rule given in §6.2, the correct TF for the oscillator equation

$$\frac{\mathrm{d}^2 u}{\mathrm{d}t^2}+cu^3=0$$

is

$$u^*=u_1 \cos \omega t$$

since the restoring term cu^3 is an odd function of u.

(a) Substitute and find the residual equation.

(b) Express the LHS in terms of multiple angles. [Appendix 6]

(c) Use harmonic balance to find the design formulae for the angular velocity ω and for the frequency f of the oscillations.

(This equation describes the oscillations of a ball in a bent tube discussed in §1.2, where the results for ω and f have been given. It has also been less accurately solved using a parabolic TF in problem 5.3.)

6.3

The equation for a large-angle pendulum (see §5.5) is

$$\frac{\mathrm{d}^2\theta}{\mathrm{d}t^2}+\frac{g}{l}\sin \theta=0.$$

The appropriate cosine TF is

$$\theta^*=\theta_0 \cos \omega t.$$

Substitution gives the residual equation

$$-\omega^2\theta_0 \cos \omega t+\frac{g}{l}\sin(\theta_0 \cos \omega t)=\mathscr{R}.$$

Although it is possible to express $\sin(\theta_0 \cos \omega t)$ as a series of multiple-angle terms, the expression involves Bessel functions and will not be used in this problem. In the absence of a suitable harmonic series, harmonic balance cannot be used, but collocation can be tried.

The qualitative sketch is shown in figures 5.7 and 5.8. The symmetry arguments of §5.5 applied to the first quarter cycle suggest collocation at one-third of the way from the extremum to the first zero. The first zero of $\cos \omega t$ is at $\omega t=\pi/2$, so the collocation point is at $\omega t=\pi/6$, where $\cos \omega t=(\sqrt{3})/2$.

(a) Collocate the residual equation at $\cos \omega t=(\sqrt{3})/2$ and hence derive the design formula

$$\omega^2=\left(\frac{g}{l}\right)\frac{\sin[(\sqrt{3})\theta_0/2]}{(\sqrt{3})\theta_0/2}.$$

(b) By expanding $\sin[(\sqrt{3})\theta_0/2]$ as far as the θ_0^3 term [appendix 2], show that the result for ω is identical to that obtained for moderate angles using harmonic balance (equation (6.11)); this result is exact to second order in θ.

(c) Using the result in (a), derive the formula for the period T, and show that it tends to the exact result for very small values of θ_0, but predicts an infinite period at $\theta_0=208°$, instead of at $\theta_0=180°$.

(This example shows that collocation according to the 'one-third of the way across' rule gives the same results as harmonic balance for moderate angles: when both processes can be used they will be found to be equivalent in this way. It also shows that, provided the solution curve is reasonably close to a sinusoid (as it is for θ_0 small), the sinusoidal TF can give exact results, and is in this sense superior to the parabolic TF for calculating periods and frequencies. For very large angles, however, when the solution curve is far from sinusoidal, the results are somewhat worse than those obtained with a parabolic TF in §5.5.)

6.4

The governing equation for the oscillations of a load of mass m hung on a light, 'hard' spring is

$$m\frac{d^2u}{dt^2}=-(\alpha u+\beta u^3)+mg$$

[mass × acceleration = elastic force + gravity force]

where u is the displacement from the equilibrium position.

(a) Write down the appropriate TF. [§6.2]

(b) Show that

$$u^{*3}=u_0^3+3u_0^2u_1\cos\omega t+3u_0u_1^2\cos^2\omega t+u_1^3\cos^3\omega t.$$

(c) Express u^{*3} in terms of multiple angles, and show that when only the constant and $\cos \omega t$ terms are retained it simplifies to

$$u^{*3}=(u_0^3+\tfrac{3}{2}u_0u_1^2)+(3u_0^2u_1+\tfrac{3}{4}u_1^3)\cos \omega t.$$

(d) Substitute u^* and the simplified form of u^{*3} obtained in (c) to find the residual equation, which will be correct except for the neglected higher harmonics.

(e) Show that harmonic balance gives

$$\alpha u_0+\beta(u_0^3+\tfrac{3}{2}u_0u_1^2)=mg$$

and

$$-m\omega^2+\alpha+\beta(3u_0^2+\tfrac{3}{4}u_1^2)=0.$$

(f) Write the first equation as

$$u_0=\frac{mg}{\alpha}-\frac{\beta}{\alpha}(u_0^3+\tfrac{3}{2}u_0u_1^2).$$

Substitute this into the second equation, and hence show that for β small

$$\omega=\left(\frac{\alpha}{m}\right)^{1/2}\left\{1+\frac{\beta}{\alpha}\left[\frac{3}{2}\left(\frac{mg}{\alpha}\right)^2+\frac{3}{8}u_1^2\right]\right\}.$$

(For a linear spring $\beta=0$, so $\omega=(\alpha/m)^{1/2}$, which does not depend on the mass m. For nonlinear springs $\beta\neq 0$, and besides the usual correction associated with the amplitude u there is also a correction associated with m.)

6.5

The governing equation for the motion of a pendulum bob of mass m subject to gravity and also to an additional constant horizontal force F is

$$\frac{d^2\theta}{dt^2}+\left(\frac{g}{l}\right)\sin\theta=\frac{F}{ml}\cos\theta.$$

(a) By expanding $\sin\theta$ and $\cos\theta$ as far as θ^3 terms, find the equation for moderate angles. [Appendix 6]

(b) Write down the appropriate sinusoidal TF.

(c) Find simplified expressions for θ^{*2} and θ^{*3}, by expressing them in terms of multiple angles, retaining only constant and $\cos \omega t$ terms. [See problem 6.5(c)]

(d) Use these expressions and harmonic balance to derive the equations

$$\frac{g}{l}(\theta_0-\tfrac{1}{6}\theta_0^3-\tfrac{1}{4}\theta_0\theta_1^2)=\frac{F}{ml}(1-\tfrac{1}{2}\theta_0^2-\tfrac{1}{4}\theta_1^2)$$

and

$$-\omega^2+\frac{g}{l}(1-\tfrac{1}{2}\theta_0^2-\tfrac{1}{8}\theta_1^2)=-\frac{F}{ml}\theta_0.$$

(e) For F, θ_0 and θ_1 all small, the first equation reduces to

$$\theta_0 = F/mg.$$

Substitute this into the second equation, and hence find a design formula for ω.

(The result shows that the frequency of a pendulum is slightly affected by a horizontal force, and this is confirmed experimentally. If the corresponding linear equation

$$\frac{d^2\theta}{dt^2} + \left(\frac{g}{l}\right)\theta = \frac{F}{ml}$$

is solved, the 'exact' result is

$$\omega^2 = g/l$$

suggesting erroneously that the horizontal force has no effect upon the frequency. This again illustrates the point made in §6.1, that linear equations are often inadequate descriptions of real oscillators.)

6.6

The cosine TF can be extended to deal with examples of equation (6.2) for which the damping term is not zero, but is small. To illustrate the technique, consider first the damped linear harmonic oscillator equation

$$\frac{d^2\theta}{dt^2} + p\frac{d\theta}{dt} + \omega_0^2\theta = 0$$

whose exact solution (2.22) is well known. The damping coefficient p will be assumed to be small.

The qualitative sketch (figure 6.2) suggests the TF

$$\theta^* = A \cos \omega t$$

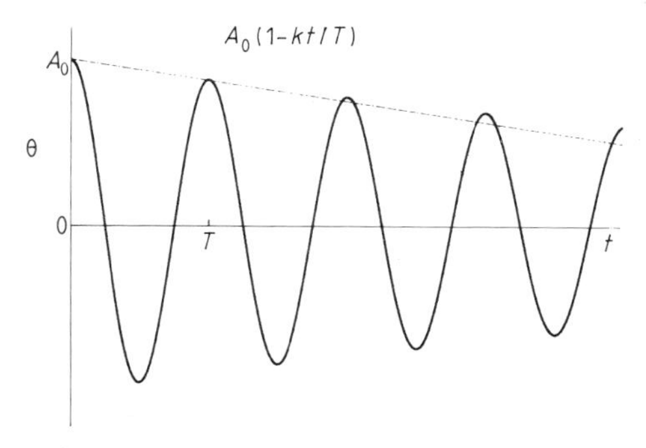

Figure 6.2 Qualitative sketch for a weakly damped free oscillator. The amplitude decays approximately linearly over one cycle.

where A is not constant, but declines from A_0 to $A_0(1-k)$ during a cycle. For p small, k will also be small, and over one cycle the decay curve of A is approximately the straight line

$$A = A_0(1 - kt/T).$$

The TF is therefore

$$\theta^* = A_0(1 - kt/T)\cos\omega t.$$

The goal is to find the constant k, which is

$$k = \frac{\text{decay of amplitude in one cycle}}{\text{amplitude at start of cycle}}.$$

(a) Find $d\theta^*/dt$ and $d^2\theta^*/dt^2$.

(b) Show that, for p small and k small (so terms in pk can be neglected), the damping term $p\, d\theta^*/dt$ is $-pA_0\omega \sin\omega t$ to first order in p.[§2.8]

(c) Using this approximation, form the residual equation and show that harmonic balance of the $\cos\omega t$ terms gives

$$-A_0\omega^2 + A_0\omega_0^2 = 0$$

whence $\omega = \omega_0$ to first order in p.

(This is the exact result (2.23) approximated to first order by neglecting the p^2 term.)

(d) Show that balancing the $\sin\omega t$ term gives

$$2A_0\frac{\omega k}{T} - pA_0\omega = 0$$

whence the required design formula is

$$k = pT/2.$$

The decay of amplitude is therefore given for short times as

$$A = A_0(1 - pt/2)$$

which agrees to first order with the exact result (2.22)

$$A = A_0\, e^{-pt/2}.$$

(The success of this TF in this linear problem suggests its application to the more difficult equation in problem 6.7.)

6.7

An important practical case of a damped oscillator equation, not exactly soluble by elementary methods, is that for a pendulum swinging through small angles, subject to air drag on the bob. It is

$$\frac{d^2\theta}{dt^2} + b\left|\frac{d\theta}{dt}\right|\frac{d\theta}{dt} + \frac{g}{l}\theta = 0.$$

(The damping force has to be written as proportional to $|\theta'|\theta'$ rather than to θ'^2 to ensure that its direction is always such as to oppose the motion†.)

It is again assumed that the damping coefficient b is small, and the TF is, as in problem 6.6,

$$\theta^* = A_0(1 - kt/T) \cos \omega t$$

where k is small and is the goal parameter.

(a) Following problem 6.6(a) and (b), show that to first order in b the damping term is

$$-bA_0^2\omega^2 |\sin \omega t| \sin \omega t.$$

(b) Using the Fourier series in appendix 6, write down the residual equation, retaining only the terms in $\sin \omega t$ and $\cos \omega t$.

(c) Noting from problem 6.6 that the formula for k is obtained from the equation found by balancing the $\sin \omega t$ term, show that

$$k = \tfrac{8}{3} b A_0.$$

(This is the required design formula, and is exact to first order in b. Unlike the linear damping case, it is independent of the period of oscillation; another difference is that the decay parameter k is no longer the same constant for every cycle, but is proportional to the amplitude at the beginning of the cycle being studied.)

6.8

In certain circumstances, the liquid in the capillary viscometer described in problem 3.7 may rise beyond the final height and then settle down in a series of decaying oscillations (figure 6.3). It is of some practical importance to establish the conditions that will ensure that this overshoot does not occur, since only then is the exponential TF used in problem 3.7 a sensible approximation. The governing equation is

$$\frac{d^2h}{dt^2} + \frac{5}{4}\left(\frac{dh}{dt}\right)^2 + bh\frac{dh}{dt} + ch = d.$$

(a) Noting that the overshoot consists of small oscillations about the final steady-state height h_∞, assume the TF

$$h^* = h_\infty + h_1 f$$

where $h_1 \ll h_\infty$ and f is a function describing the oscillations. Substitute this TF into the equation, assuming h_1 small, to derive

$$h_1 f'' + bh_\infty h_1 f' + ch_\infty + ch_1 f = d.$$

† $|\theta'|$ is the modulus of θ'; $|\theta'|$ has the same value as θ' except that its sign is always positive, regardless of whether θ' is positive or negative.

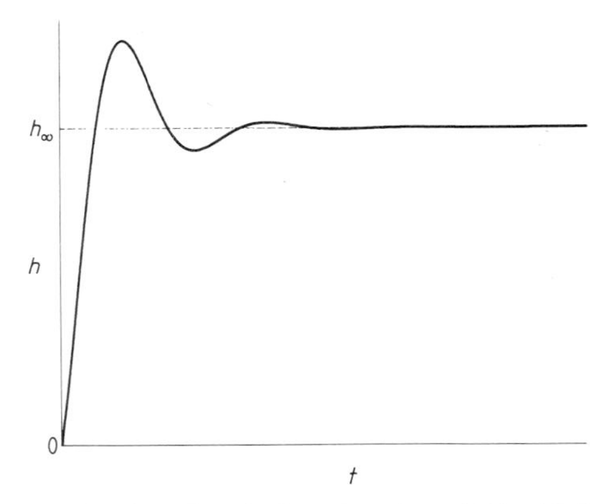

Figure 6.3 Qualitative sketch for the rise of liquid in a capillary viscometer, showing the oscillatory overshoot expected for a low-viscosity liquid.

(b) Deduce that $ch_\infty = d$, and hence that

$$f'' + bh_\infty f' + cf = 0.$$

(c) Referring to §2.5, rewrite the equation in the standard form of a damped, linear oscillator equation, and hence show that the assumed small oscillations are only possible if

$$b < 2c/h_\infty.$$

[Use the discriminant formula (2.24)]

(This must be, at least approximately, the condition for overshoot to occur. Since $b = 8\nu/R^2$, this shows that overshoot is associated with small viscosity ν and large tube radius R.)

Chapter 7

Forced oscillation and resonance

7.1 Introduction

The cosine TFS used in the preceding chapter can also be applied to forced oscillations to gain some useful preliminary insight into these problems.

The simplest equation that can be used to model forced oscillations is

$$\underset{}{\frac{d^2u}{dt^2}} + \underset{[\text{damping}]}{p\frac{du}{dt}} + \underset{[\text{restoring}]}{\omega_0^2 u} = \underset{[\text{forcing}]}{A\cos\omega t}. \tag{7.1}$$

Except for the addition of the forcing term, the equation is identical to that for the free damped harmonic oscillator, i.e.

$$\frac{d^2u}{dt^2}+p\frac{du}{dt}+\omega_0^2u=0. \tag{7.2}$$

As explained in §2.5, this equation models systems like the pendulum, which may be initially set into free oscillation at a natural frequency

$$f_f=\frac{\omega_f}{2\pi}=\frac{\omega_0}{2\pi}\left(1-\frac{p^2}{4\omega_0^2}\right)^{1/2}.$$

These oscillations, however, will die out because of the energy losses caused by friction, represented in the equation by the damping term $p\,du/dt$.

The forced oscillation equation (7.1) describes the same inherently oscillatory system, but there is now an external force acting, represented by the additional forcing term $A\cos\omega t$. The work done by this additional force can balance the energy lost by friction, with the result that the oscillations eventually settle down to a steady amplitude.

The system no longer oscillates at its natural frequency f_f, but at the frequency of the oscillating external force, so that

$$f=\frac{\omega}{2\pi}$$

where ω is the angular velocity in the forcing term $A\cos\omega t$.

A qualitative sketch of the steady response of the system to the forcing oscillation is shown in figure 7.1. The oscillations of the independent variable u have the same frequency as the forcing term, but are shifted by an unknown phase angle ϕ. The amplitude of u is also unknown. The usual purpose of mathematical models is to find formulae for the amplitude (and perhaps phase) in terms of the parameters of the physical equation. In this chapter, we shall concentrate upon finding design formulae for the amplitude u_1, so as to study its dependence upon ω, the angular velocity of the forcing oscillation.

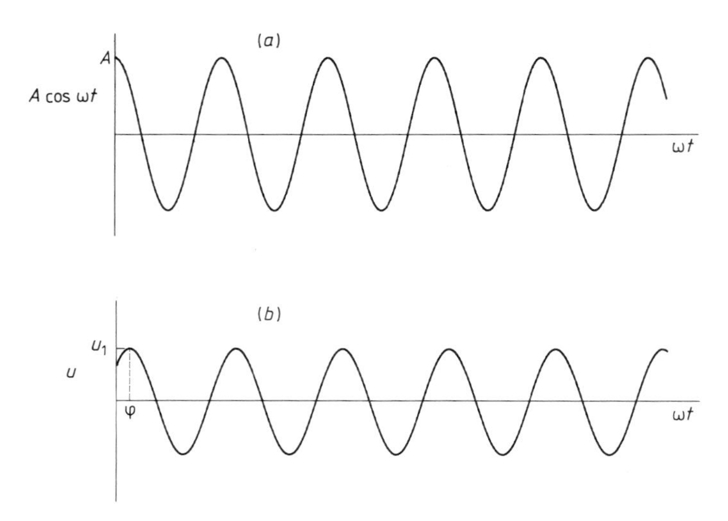

Figure 7.1 Qualitative sketch for forced oscillations of a damped harmonic oscillator: (*a*) the forcing term $A \cos \omega t$; (*b*) the response $u = u_1 \cos(\omega t - \phi)$, at the same frequency but shifted by an unknown phase angle ϕ. The origin of the timescale is arbitrary.

A graph of the variation of amplitude against ω is called a *resonance curve*, and its shape can be anticipated intuitively to look something like figure 7.2. The system has a natural tendency to oscillate at an angular velocity ω_f, and it is therefore reasonable to anticipate that the largest oscillations will be caused when the driving angular velocity ω exactly coincides with ω_f. When ω is greater or less than ω_f, it may be supposed that the oscillation amplitude will fall off, giving the typical resonance curve shown in figure 7.2. The examples will show that this intuitive picture is broadly fulfilled, but that the resonance curves for all but the simplest cases have many interesting and unexpected features, with important physical consequences.

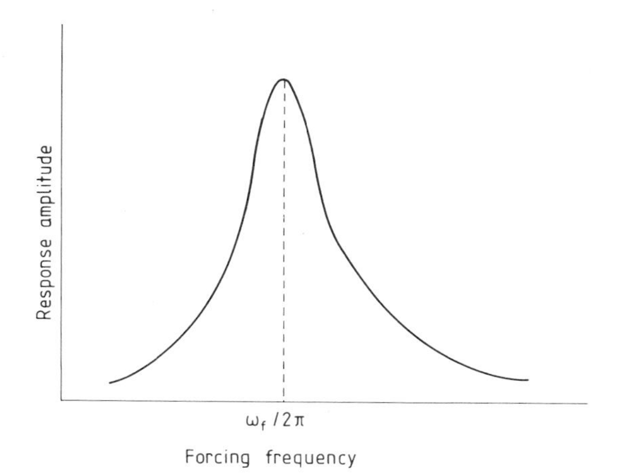

Figure 7.2 Intuitive sketch of a resonance curve illustrating the maximum response when the forcing frequency equals the natural frequency of oscillation $\omega_f/2\pi$.

7.2 The QSTF method for forced oscillations

7.2.1 The damped simple harmonic oscillator

The method for forced oscillations uses the same trial functions and harmonic balancing technique already described in Chapter 6. As a first simple example it will be used to draw the resonance curves for the damped simple harmonic oscillator already discussed in general terms in §7.1.

The equation for the displacement of the bob of a simple pendulum with linear damping and restoring terms, but with an additional forcing term, is equation (7.1), i.e.

$$\frac{d^2u}{dt^2}+p\frac{du}{dt}+\omega_0^2u=A\cos\omega t.$$

The oscillatory forcing term corresponds to a time-varying force, which could be provided, for instance, by an electromagnet supplied with an alternating current, as in a mains-operated clock. The qualitative sketch in figure 7.1 suggests the TF

$$u=u_1\cos(\omega t-\phi) \tag{7.3}$$

where the amplitude u_1 and relative phase ϕ are unknown parameters. Notice that, unlike the TF for the free oscillator, in this TF ω is not an unknown quantity, but is equal to the known angular velocity of the forcing term.

The work is simplified if, instead of incorporating the phase difference ϕ into the TF, it is transferred to the forcing term in the equation. The equation is then

$$\frac{d^2u}{dt^2}+p\frac{du}{dt}+\omega_0^2u=A\cos(\omega t+\phi)$$

$$=A(\cos\phi\cos\omega t-\sin\phi\sin\omega t) \tag{7.4}$$

and the TF is

$$u^*=u_1\cos\omega t. \tag{7.5}$$

It will be seen that this states that the forcing oscillation leads the TF oscillation by an angle ϕ, whereas equation (7.1) with TF (7.3) stated that the TF lagged the forcing term by ϕ; the two statements are physically equivalent.

7.2.2 *Solution*

Substitution of TF (7.5) into equation (7.4) gives

$$-\omega^2u_1\cos\omega t-p\omega u_1\sin\omega t+\omega_0^2u_1\cos\omega t$$

$$=A\cos\phi\cos\omega t-A\sin\phi\sin\omega t+\mathscr{R}$$

or

$$[(-\omega^2+\omega_0^2)u_1-A\cos\phi]\cos\omega t-(p\omega u_1-A\sin\phi)\sin\omega t=\mathscr{R}.$$

For two unknowns, harmonic balance requires that the coefficients of the two terms on the LHS be equated to zero, giving

$$(\omega_0^2-\omega^2)u_1=A\cos\phi \tag{7.6}$$

and

$$p\omega u_1=A\sin\phi. \tag{7.7}$$

Since there are no terms involving multiple angles, the procedure in this case makes the residual $\mathscr{R}$ zero for all values of t, implying an exact solution.

Equations (7.6) and (7.7) are two simultaneous equations for the two unknowns ϕ and u_1. Since the solution for u_1 is required, ϕ is eliminated by first squaring and then adding the two equations to give

$$(\omega_0^2-\omega^2)^2u_1^2+p^2\omega^2u_1^2=A^2(\cos^2\phi+\sin^2\phi)=A^2.$$

Solving for u_1 gives the required (exact) design formula

$$u_1=\frac{A}{[(\omega_0^2-\omega^2)^2+p^2\omega^2]^{1/2}}. \tag{7.8}$$

The resonance curves described by this function are shown in figure 7.3 for various values of the damping parameter p.

The curves show the typical peaks expected intuitively, but the height and position of the resonant peaks and the shape of the curves depend strongly upon the value of p. If there is no damping ($p=0$), the peak amplitude is infinite

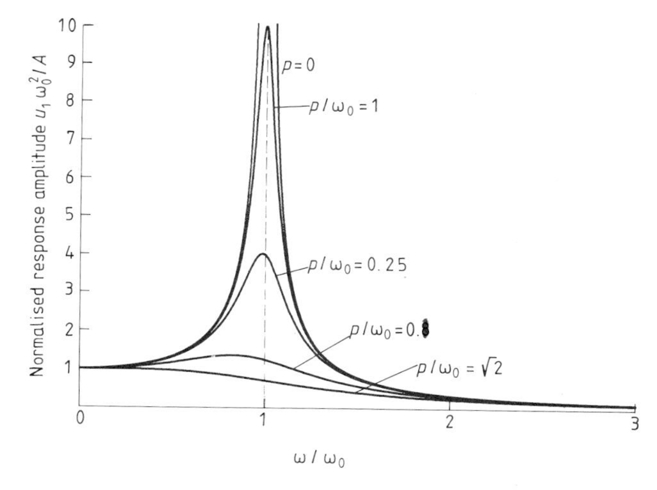

Figure 7.3 Resonance curves for a forced, damped harmonic oscillator. The curves show values of the normalised response amplitude $u_1\omega_0^2/A$, given by formula (7.8), for various values of the damping parameter p.

and occurs when $\omega = \omega_0$. For small but finite damping, the peak remains close to ω_0. As the damping increases further, the peak moves to the left and declines. Eventually the peak disappears altogether when the damping reaches the value $p = \sqrt{2}\omega_0$.

The value of frequency at the peak of the resonance curve, found by differentiating the exact formula (7.8), is given by

$$f_r = \frac{\omega_0}{2\pi}\left(1 - \frac{p^2}{2\omega_0^2}\right)^{1/2}. \tag{7.9}$$

Frequency f_r is called the *resonant frequency*, defined as the forcing frequency that produces the maximum response amplitude. It should be noted that the resonant frequency f_r is not exactly equal to the natural frequency f_f, though it is very close except for heavy damping.

7.3 Primary resonance of a weakly nonlinear oscillator: the Duffing equation

Mechanical systems often display cubic nonlinearity in the restoring term, as discussed in §6.3. If an oscillatory forcing term is applied to such a system, having a frequency close to that of free oscillations, then resonance is induced in much the same way as for the linear system. The governing equation for such an oscillator is the *Duffing equation*

$$\frac{d^2u}{dt^2}+p\frac{du}{dt}+\omega_0^2u+\gamma u^3=A\cos\omega t. \tag{7.10}$$

This differs from the linear simple harmonic forced oscillator equation (7.1) only in the addition of the nonlinear term γu^3. This nonlinearity, even if γ is small, produces some remarkable changes in the nature of the solution and the resonance curves.

The primary goal is to construct approximate resonance curves for these nonlinear systems, by finding a design formula for the amplitude of the steady-state response which shows its dependence upon the frequency $\omega/2\pi$ of the forcing function.

The solution procedure follows the method used to treat the forced linear case in §7.2.

The qualitative sketch is the same as for the linear case, as shown in figure 7.1. The TF is also the same, and again it greatly simplifies the working to transfer the phase factor to the forcing term; accordingly we write the equation as

$$\frac{d^2u}{dt^2}+p\frac{du}{dt}+\omega_0^2u+\gamma u^3=A\cos(\omega t+\phi) \tag{7.11}$$

and use the TF

$$u^*=u_1\cos\omega t. \tag{7.12}$$

The unknown parameters are u_1 and ϕ, and the goal is to find a design formula for the response amplitude u_1.

Substituting the TF (7.12) into the equation (7.11) and making the substitution

$$\cos^3\omega t=\tfrac{1}{4}(3\cos\omega t+\cos 3\omega t)$$

gives

$$-\omega^2u_1\cos\omega t-p\omega u_1\sin\omega t+\omega_0^2u_1\cos\omega t+\tfrac{3}{4}\gamma u_1^3\cos\omega t+\tfrac{1}{4}\gamma u_1^3\cos 3\omega t$$
$$=A\cos\phi\cos\omega t-A\sin\phi\sin\omega t+\mathscr{R}.$$

Collecting terms in $\cos\omega t$, $\sin\omega t$ and $\cos 3\omega t$ yields the residual equation

$$[(-\omega^2+\omega_0^2+\tfrac{3}{4}\gamma u_1^2)u_1-A\cos\phi]\cos\omega t$$
$$+(-p\omega u_1+A\sin\phi)\sin\omega t+\tfrac{1}{4}\gamma u_1^3\cos 3\omega t=\mathscr{R}. \tag{7.13}$$

Unlike the linear case, the residual equation cannot be balanced exactly for all values of t. The solution obtained will therefore be approximate. Harmonic balance by equating the coefficients of $\cos\omega t$ and $\sin\omega t$ separately to zero gives two equations for the two unknowns u_1 and ϕ. Thus,

$$(\omega_0^2-\omega^2+\tfrac{3}{4}\gamma u_1^2)u_1/A=\cos\phi \tag{7.14}$$

and

$$p\omega u_1/A = \sin\phi. \tag{7.15}$$

By squaring and adding these two equations ϕ is eliminated, since $\cos^2\phi + \sin^2\phi = 1$; then

$$[(\omega_0^2 - \omega^2 + \tfrac{3}{4}\gamma u_1^2)^2 + p^2\omega^2](u_1/A)^2 = 1.$$

Solving for u_1 gives the design formula for the response amplitude:

$$u_1 = \frac{A}{[(\omega_0^2 - \omega^2 + \frac{3}{4}\gamma u_1^2)^2 + p^2\omega^2]^{1/2}}. \tag{7.16}$$

As a check it may be observed that when $\gamma = 0$ (no nonlinearity in the modelling equation) this design formula reduces to the result (7.8) already found for the linear case.

7.4 Nonlinear resonance

The design formula (7.16) is not an explicit relationship for $u_1(\omega)$, since u_1 occurs on both sides of the equation. Attempts to solve directly for $u_1(\omega)$ are not recommended because the equation is cubic in u_1^2. Instead the method of inverse tabulation explained in appendix 4 should be used to solve for $\omega(u_1)$. This only involves solving a quadratic equation for ω^2, using the usual formula. The results are shown in figure 7.4(*a*) and (*b*) for selected values of the parameters A, p and γ.

Figure 7.4(*b*) shows clearly the effect of the nonlinearity. As γ increases from zero, the resonance tilts away from the vertical. Positive values of γ tilt the curve to the right; negative values tilt it to the left. The effect is not symmetrical, as may be seen by comparing the curves for $\gamma = \pm 0.02$. This tilting, which is absent in the linear resonance curves like figure 7.3, has practical consequences which could not be anticipated if the linear equations were used as mathematical models.

Consider the typical nonlinear resonance curve shown by the full curve in figure 7.5, and imagine an experiment in which the frequency of the forcing oscillation applied to the system is initially very small. The response amplitude is then given by the point A on the curve. As the forcing frequency increases, the response amplitude follows the curve through point B until eventually it reaches the peak at C. If the frequency is increased very slightly, the only possible value of u_1 is at point D on the curve, so the amplitude must jump from a high value to a low value, as shown by the longer broken line. This jump in amplitude is a type of instability associated with nonlinear resonance but absent from linear resonance. Further increase in the frequency beyond point D produces no new behaviour, the response amplitude falling steadily through points E and F.

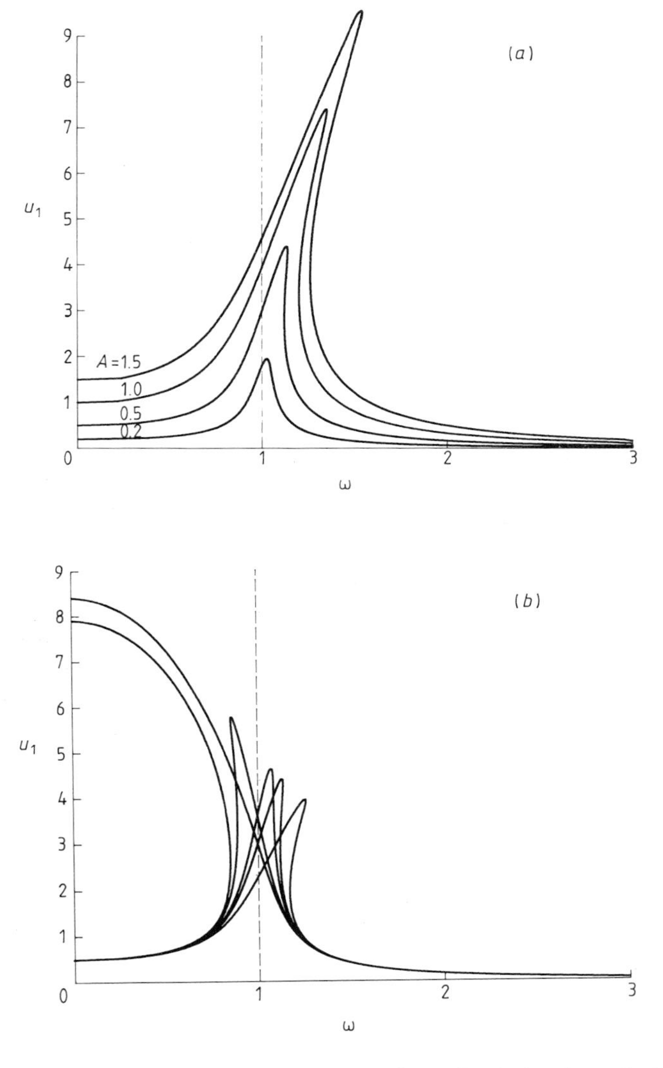

Figure 7.4 Nonlinear resonance curves from the design formula (7.16). (*a*) The effect of forcing amplitude A ($\gamma = 0.02$, $p = 0.1$). (*b*) The effect of nonlinearity γ ($A = 0.5$, $p = 0.1$). $\omega_0 = 1$ for all curves.

Now consider the effect of reducing the frequency, starting at point F. At first, the amplitude rises steadily through the points E and D until eventually the point G is reached. Further reduction in frequency forces the response to jump to the upper curve at point H. Once on the upper curve, the response

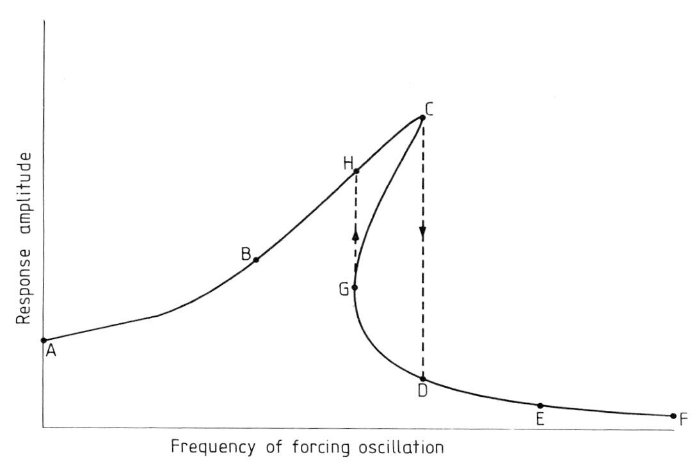

Figure 7.5 Amplitude instability caused by tilting of the resonance curve.

amplitude falls steadily through point B, back to the starting point A at zero frequency.

The path followed by the response amplitude is redrawn in figure 7.6. This shows clearly the phenomenon of *hysteresis*; that is, the amplitude is not uniquely determined by the forcing frequency, but depends on the past history of the system. Note that practical oscillators can never operate on the portion of the resonance curve between points C and G in figure 7.5.

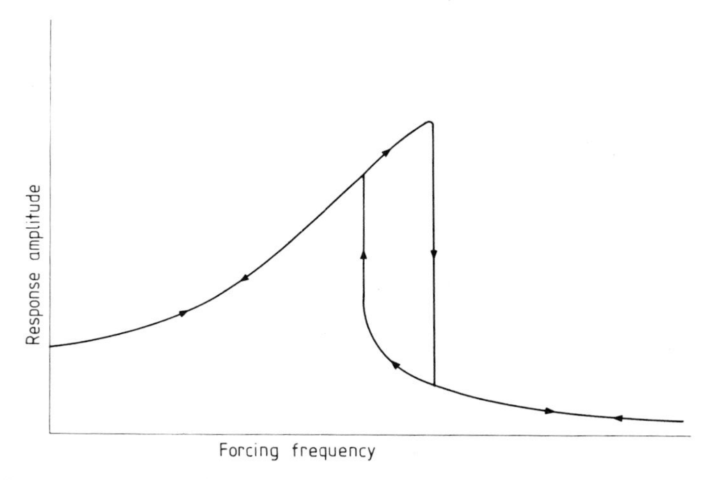

Figure 7.6 Hysteresis in the amplitude of a forced nonlinear oscillator. The amplitude follows a hysteresis loop as the frequency is cycled up and down through resonance.

These features of the behaviour of forced nonlinear oscillators are examples of a wide range of interesting, unexpected and often important phenomena that have no analogue in the linear case (see, for example, problem 7.3). The interested reader will find a comprehensive account of nonlinear resonance in Nayfeh and Mook (1979).

Problems

7.1

A capacitor microphone consists of a capacitor with one of its plates free to move in response to sound pressure waves in the surrounding air. The effect is to vary the effective spacing between the plates by a small fraction ε of the equilibrium value. The capacitance varies accordingly, and its value may be expressed in the form

$$C = C_0(1 - \varepsilon \cos \omega t).$$

In use, the microphone is connected in series with a resistor R and a DC voltage V_0; the signal voltage v is taken across R.

The governing equation for the charge Q on the capacitor as a function of time, in response to a sound wave of angular velocity ω, is

$$RC_0 \frac{\mathrm{d}Q}{\mathrm{d}t} + Q(1 - \varepsilon \cos \omega t) = C_0 V_0.$$

The immediate goal is to find a design formula for the amplitude of the charge oscillations; the signal voltage can then be deduced.

The DC voltage V_0 imposes a bias in the circuit, so a shifted cosine TF must be used. Also there may be a phase difference between the applied sound wave and the resulting charge response, and as in §§7.2 and 7.3 it is simpler to incorporate the phase angle in the equation than in the TF.

(a) Substitute the TF

$$Q^* = Q_0 + Q_1 \cos \omega t$$

in the equation

$$RC_0 \frac{\mathrm{d}Q}{\mathrm{d}t} + Q[1 - \varepsilon \cos(\omega t + \phi)] = C_0 V_0$$

to obtain the residual equation.

(b) There are three unknown constants: the goal quantity Q_1, the DC charge Q_0 and the phase angle ϕ. Three terms in the residual equation must therefore be balanced. These are the constant term and the two fundamental terms containing $\sin \omega t$ and $\cos \omega t$. Write down the three equations that result from balancing these three terms.

(c) Use the two equations found by balancing the fundamental terms to obtain expressions for $\cos\phi$ and $\sin\phi$. Use the identity $\cos^2\phi+\sin^2\phi=1$ to eliminate ϕ, and hence obtain the relation

$$Q_1=\frac{\varepsilon Q_0}{(1+\omega^2R^2C_0^2)^{1/2}}.$$

(d) Use the equation found by balancing the constant term to eliminate Q_0, and show that for small ε, the design formula for Q_1 is the result in (c), with C_0V_0 in place of Q_0.

(e) Use the relations $v=iR$ and $i=\mathrm{d}Q_1/\mathrm{d}t$ to estimate the signal voltage.

(f) (This part should only be attempted if the optional section §6.7 has been studied.) Verify that the residual ratio is $\varepsilon/2$. [§6.7] (Since ε has been taken as small, this confirms that the design formula is numerically useful. The minimum residual is of some practical significance in this problem, as it is a measure of the harmonic distortion introduced by the microphone.)

7.2

In addition to primary resonance, it is possible for nonlinear oscillators to display resonances at lower frequencies than that of the forcing term. Since these normally occur at frequencies that are integral submultiples of the forcing frequency, they are called *subharmonic resonances*. The simplest system that demonstrates this possibility is modelled by the equation

$$\frac{\mathrm{d}^2u}{\mathrm{d}t^2}+u^3=A\cos 3\omega t.$$

(a) Substitute the TF

$$u^*=u_1\cos\omega t$$

which describes a response at one-third of the forcing frequency and obtain the residual equation.

(b) Show that the residual equation can be balanced exactly by making

$$\omega=3^{1/2}u_1/2$$

and

$$u_1=(4A)^{1/3}.$$

Elimination of u_1 gives

$$\omega=\frac{3^{1/2}}{2^{1/3}}A^{1/3}.$$

If this condition is satisfied, the TF is an exact solution of the equation.

(If ω and A happen by coincidence to satisfy the condition, the system will suffer a forced oscillation at one-third of the forcing frequency, even though the response at the forcing frequency is zero. In real oscillators, it is unlikely that the primary response would be totally absent; nevertheless this subharmonic resonance can be a strong effect. There is a well attested case of an aircraft being destroyed as a consequence of it.)

7.3

An important class of practical devices is so-called *parametric oscillators*. In these, one of the parameters of the device is varied periodically (or *pumped*) in such a way as to produce oscillations at a lower frequency. An everyday example is a child's swing, which is effectively a pendulum whose length is varied by the movement of the child's body. Observation of a swing shows that the amplitude builds up most efficiently when the swing is pumped at or near twice the natural frequency of free oscillations of the swing.

The governing equation for small amplitudes is the *Mathieu equation*

$$\frac{d^2u}{dt^2}+\omega_0^2(1+\varepsilon\cos\Omega t)u=0.$$

The term $\varepsilon\cos\Omega t$ represents the variation in the restoring force, caused by varying one of the parameters (such as the length of the swing) at angular velocity Ω. It is assumed that ε is a small parameter.

The goal is to find what values of Ω cause the oscillation amplitude to build up. The experimental observations referred to above suggest that Ω must be close to but not necessarily identical to $2\omega_0$, so it is helpful to write $\Omega=2\omega$, and to look for oscillations with angular velocity ω near ω_0. The method is similar to that followed in problems 6.6 and 6.7, where the slowly changing amplitude was modelled by a linear function of time. In this case, we are looking for an increase, rather than a decrease, in the amplitude, so the TF is

$$u^*=A_0(1+kt/T)\cos\omega t$$

where the plus sign corresponds to the increase in amplitude.

(a) Substitute the TF into the equation, rewritten in the following way:

$$\frac{d^2u}{dt^2}+\omega_0^2[1+\varepsilon\cos(2\omega t+\phi)]u=0$$

where the phase angle ϕ has again been incorporated in the equation, rather than in the TF. Find the residual equation, retaining only the fundamental terms.

(b) Show that the $\cos\omega t$ term can be balanced exactly by making $\omega^2-\omega_0^2=\frac{1}{2}\varepsilon\omega_0^2\cos\phi$.

(c) The $\sin\omega t$ term cannot be balanced exactly for all t, but the out-of-balance part contains εk, which is of second order of smallness. Confirm that first-order balance can be obtained by making

$$-2\omega k/T=\tfrac{1}{2}\varepsilon\omega_0^2\sin\phi.$$

(d) Eliminate ϕ by using the identity $\cos^2\phi+\sin^2\phi=1$, to obtain the following equation for k:

$$k^2=\frac{\varepsilon^2\omega_0^4-4(\omega^2-\omega_0^2)^2}{16\omega^2}T^2.$$

(e) For real values of k, the numerator must be $\geqslant 0$. Show that for ε small, this leads to the result

$$2\omega_0(1-\tfrac{1}{4}\varepsilon)\leqslant\Omega\leqslant 2\omega_0(1+\tfrac{1}{4}\varepsilon).$$

(This is a design formula for the range of pump frequencies that will cause the oscillations to build up. It shows that if $\Omega=2\omega$ lies within a small band of frequencies whose limits are $\varepsilon\omega_0/2$ on either side of $2\omega_0$, the amplitude can increase. This is the correct result to first order in ε.)

Chapter 8

Exact solution of partial differential equations

8.1 Introduction

The chapters that follow extend the QSTF methods to find approximate solutions of some important partial differential equations. The merits and limitations of the QSTF method for PDES cannot be appreciated without a preliminary understanding of their exact solutions, and the present chapter reviews these to an extent sufficient for an understanding of the later chapters.

Of the PDES that arise in the experimental sciences, three types are of particular importance. They are the Laplace equation, the conduction/diffusion equation and the wave equation. The QSTF method is most useful for the latter two equations, and it is the exact solutions of these that are discussed in this chapter; the examples treated by the QSTF method in the following chapters are all of these types.

(a) *Conduction/diffusion equations*, which occur commonly as the mathematical models in heat conduction and diffusion problems, are characterised by the Laplacian operator ∇^2 (see appendix 5) and the first partial differential $\partial/\partial t$ with respect to time, no higher time differential being present. It is sufficient for present purposes to consider the common one-dimensional form

$$D\frac{\partial^2 u}{\partial x^2}+G=\frac{\partial u}{\partial t} \tag{8.1}$$

where D and G are constants. Conduction/diffusion equations usually describe smooth changes with time, the first-order differential $\partial u/\partial t$ corresponding to the ordinary differential $\mathrm{d}u/\mathrm{d}t$ in the time-dependent ODE 1 studied in Chapter 3.

(b) *Wave equations*, which describe free oscillations, are characterised by the presence of the Laplacian operator ∇^2, together with the second time differential $\partial^2/\partial t^2$. The most familiar examples have the one-dimensional form

$$\frac{\partial^2 u}{\partial x^2}=\frac{1}{c^2}\frac{\partial^2 u}{\partial t^2} \tag{8.2}$$

where c is a constant (the wave velocity).

The second-order differential $\partial^2 u/\partial t^2$ corresponds to d^2u/dt^2 in the inherently oscillatory equations studied in Chapters 6 and 7.

The solution of both of these types of PDE will be a function $u(x, t)$ of both the spatial coordinate x and the time t. The classical *separation technique* for solving both these types of PDE is to assume that $u(x, t)$ can be written in the form

$$u(x, t) = XT$$

where X is a function of x only, and T is a function of t only. This leads to two ordinary differential equations, one for T and one for X. The equation for X derived by 'separation' of equation (8.2), for instance, is:

$$\frac{d^2X}{dx^2} + k^2X = 0 \tag{8.3}$$

where k^2 is called the separation constant. This equation is mathematically of the same form as the equation (2.14) for simple harmonic oscillations such as those of a simple pendulum; however, because the independent variable x is in space and not in time, the solution will have to satisfy boundary conditions instead of initial conditions. It will be shown in §8.3 that there is no longer a unique solution, but a whole set of possible solutions, called *eigenfunctions*, which satisfy both the equation and its boundary conditions. There is, therefore, a significant difference from the inherently oscillatory equations in time, studied in Chapters 2 and 6, which have only one solution satisfying the two initial conditions.

An ODE 2 like equation (8.3) which has eigenfunction solutions is called an *eigenvalue equation.*

The exact solution of linear PDES therefore involves two steps; first, a separation into two or more ODES, and then the solution of an eigenvalue equation in terms of its eigenfunctions. So PDES, eigenvalue equations and eigenfunctions are therefore closely related. The two examples described in the following sections will clarify this relationship, and will be frequently referred to in the later chapters.

8.2 Separation of wave equations: the vibrating string

It has already been mentioned that the classical technique for separating PDES is to express the solution as a product of functions each depending upon only one variable. Wave equations, which we shall consider first, arise in mathematical models of practical systems that describe oscillations in time; this allows a more direct method of separation to be used.

A familiar example is the simplified model for the vibrating string of a guitar. When such a string is plucked, its initial shape is like that shown by situation A

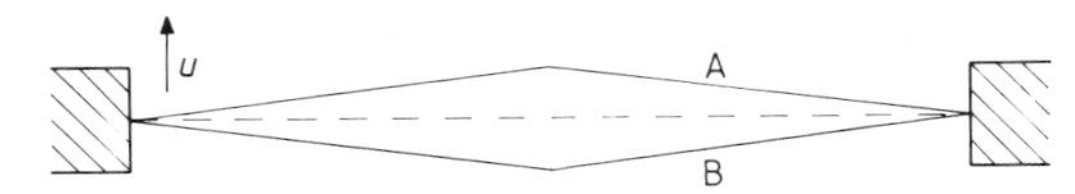

Figure 8.1 Displacement u of a plucked guitar string, showing initial displacement (A) and displacement after half a cycle of vibration (B).

of figure 8.1. It then vibrates so that after half a cycle the displacement u is shown by situation B of figure 8.1; and after a full cycle the string has returned to its original position, A.

The governing equation (ignoring damping) is equation (8.2), but the physical content is clearer if the two sides are changed over. It then more directly describes the transverse force balance on a short element δl of the string:

$$\frac{\partial^2 u}{\partial t^2} = c^2 \frac{\partial^2 u}{\partial x^2}. \tag{8.4}$$

$$\left[\begin{matrix}\text{acceleration} \\ \text{of } \delta i\end{matrix} \propto \begin{matrix}\text{force} \\ \text{on } \delta l\end{matrix}\right]$$

The presence of the differentials $\partial^2/\partial t^2$ and $\partial^2/\partial x^2$, as well as the knowledge that the equation models an oscillation, identifies equation (8.4) as a wave equation.

The purpose of the following discussion is to revise the separation technique, and in particular to remind the reader how the corresponding eigenvalue problem arises.

When it is known from the physical background to the PDE that the solution is a free oscillation in time, it is simplest to separate the equation by starting with the TF

$$u = X \cos \omega t \tag{8.5}$$

where X is a function of x only.

Substitution of the TF (8.5) into the equation (8.4) gives

$$X(-\omega^2 \cos \omega t) = c^2 \frac{d^2 X}{dx^2} \cos \omega t.$$

Dividing by $\cos \omega t$ and rearranging gives

$$\frac{d^2 X}{dx^2} + \left(\frac{\omega^2}{c^2}\right) X = 0. \tag{8.6}$$

This is an eigenvalue equation, identified as such because (1) it is an inherently oscillatory ODE 2, (2) the oscillations depend upon the spatial variable x and (3) the solution must be fitted between boundary conditions. The angular velocity ω, whose value is the practical goal of the solution, appears in this equation as an unknown constant. The problem has therefore been reduced to finding the solution of equation (8.6) in the hope that this will reveal the value of ω.

8.3 Spatial eigenfunctions and eigenvalues for a vibrating string

The general solution of equation (8.6) is (§2.4)

$$X = A \sin \frac{\omega}{c} x + B \cos \frac{\omega}{c} x.$$

This solution must be made to fit the boundary conditions for X, which derive from the conditions that the displacement u must always be zero at the ends of the string, regardless of the time. Thus X itself must be zero at the ends, so the conditions are

$$X(0) = 0 \qquad X(L) = 0.$$

The first of these is satisfied by making B zero, so the solution simplifies to

$$X = A \sin \frac{\omega}{c} x \tag{8.7}$$

which is zero at $x = 0$.

Figure 8.2(*a*)–(*c*) shows that only certain sine waves can satisfy the second boundary condition, $X(L) = 0$. The functions shown are the first three members of the set of functions

$$X_n = A \sin \frac{n\pi x}{L} \qquad n = 1, 2, \ldots. \tag{8.8}$$

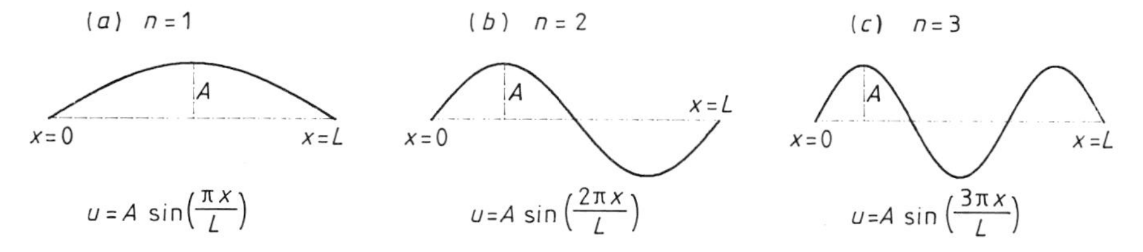

Figure 8.2 The first three functions that satisfy equation (8.6) and the boundary conditions $X(0) = 0$ and $X(L) = 0$. They are the functions (8.8) for (*a*) $n = 1$, (*b*) $n = 2$ and (*c*) $n = 3$.

The only way in which both the equation and its boundary condition can be satisfied is if the solution (8.7) is identical with one of the set of functions (8.8). Comparing the two shows that this is true if

$$\omega/c = n\pi/L$$

so that the possible values of ω are given by

$$\omega_n = n\pi c/L \qquad n = 1, 2, \ldots. \tag{8.9}$$

The eigenvalue equation can be solved consistently with the boundary conditions only if ω has one of these special values, ω_n, which are called the eigenvalues of ω.

Provided ω is equal to one of these eigenvalues, the equation can be solved, and the corresponding solutions are called the *spatial eigenfunctions*. For the vibrating string example, the spatial eigenfunctions are the set of functions X_n defined in the solution (8.8), since X_n is a solution of the equation when $\omega = \omega_n$. If ω is not one of the eigenvalues, the equation has no solution that satisfies the boundary conditions.

At this stage the practical goal of finding the value of ω has been reached. The answer is that ω does not have a unique value, but is one of a set of possible values given by the formula (8.9). The lowest frequency emitted by a vibrating string, which is the one which the ear hears most clearly and therefore decides the effective musical frequency or pitch, corresponds to the lowest value of n ($n = 1$), and is therefore

$$f = \omega_1/2\pi = c/2L. \tag{8.10}$$

The frequencies corresponding to larger values of n are harmonic frequencies which correspond musically to overtones of the fundamental note. Their presence affects the timbre but not the pitch of the note.

In many practical problems, including the present example, the prime goal is to find the lowest eigenvalue. There is no great physical interest in completing the solution to find the form of the function $u(x, t)$. However, the complete solutions will be studied a little further in order to help to understand the TF methods to be used later, and to show how the TF results relate to the exact solutions.

8.4 Normal modes of a vibrating string

The solution of the original PDE (8.4) can be found by returning to the original TF which was used for its separation. The TF (8.5) is

$$u_n(x, t) = X_n \cos \omega_n t. \tag{8.11}$$

The suffices have been added, since it is now known that X and ω are not unique, but are members of the sets X_n and ω_n. The corresponding possible functions u_n are found by substituting the results (8.8) and (8.9) for X_n and ω_n into the TF. This gives

$$u_n(x, t) = A_n \sin\left(\frac{n\pi}{L} x\right) \cos \omega_n t \qquad n = 1, 2, \ldots \tag{8.12}$$

where $\omega_n = n\pi c/L$ ($n = 1, 2, \ldots$).

Thus there is a whole set of possible functions which are the solutions

$u(x, t)$ of the original PDE, and u_n is called the nth eigenfunction of this PDE.

The eigenfunction solutions (8.12) just derived are mathematical representations of what are called the *normal modes* of vibration of a taut string. The lowest-frequency vibration, having $n=1$, is the first or *fundamental mode*, while the vibrations for $n=2, 3, \ldots$ correspond to the second and higher modes. Any of these modes of vibration are theoretically possible, since their corresponding eigenfunctions are valid solutions; in practice, however, it is almost impossible to induce a string to vibrate purely in one of these normal modes.

The reason becomes clear if t is put equal to zero in the formula (8.11); since for $t=0$, $\cos \omega_n t = 1$ for all values of ω_n, the result is that

$$u_n(x, 0) = X_n. \tag{8.13}$$

Thus, to excite the nth normal mode the initial shape of the string (when $t=0$) must be equal to X_n.

The shapes of the spatial eigenfunction X_n for the three lowest modes have already been shown in figure 8.2. It is obviously very difficult to pull out a guitar string to any of these smooth shapes, and this is the reason that it is practically impossible to excite vibrations corresponding exactly to one of the normal modes.

A more realistic starting shape is shown in figure 8.1, in which the string is plucked at the centre and then released at $t=0$, so that the starting shape, $u(x, 0)$, is a triangle. The technique for matching such an initial shape is based on the superposition principle; this states that, as the PDE is linear, any of its solutions can be linearly combined and this combination will itself be a solution. A very general solution is therefore

$$u(x, t) = A_1 u_1 + A_2 u_2 + A_3 u_3 + \cdots \tag{8.14}$$

where $u_1, u_2, \ldots$ are the eigenfunctions for the first, second, ... modes, and $A_1, A_2, \ldots$ are arbitrary coefficients which must be chosen to satisfy the initial conditions. This involves using Fourier series, and need not be considered here.

The typical vibration of a string does not, therefore, correspond to any individual normal mode, but is made up of a mixture of these modes; the amplitudes $A_1, A_2, \ldots$ of each mode depend upon the way the string is plucked, which defines the string's shape at $t=0$.

It is, however, possible to attach a meaning to each term of the series (8.14). The first term with $n=1$ corresponds to the fundamental mode and frequency; the next term, with $n=2$, to the second mode and second harmonic; and the next to the third mode and third harmonic; and so on. The relative amplitudes of the fundamental and harmonics are given by the coefficients A_1, A_2, A_3, and usually these amplitudes diminish rapidly as n increases. The fundamental mode not only has the lowest frequency, but it is also very much the loudest, both facts explaining its overriding musical importance.

To summarise and generalise from the example, the exact solution of a wave equation will normally be a relatively complicated series, made up by the linear combination of the eigenfunctions of the equation. The most important term physically is the first, which corresponds to the fundamental mode of oscillation. Although in some theoretical problems, notably in quantum physics, knowledge and understanding of the higher eigenfunctions is of crucial importance, for many practical problems only the first eigenfunction, which corresponds to the fundamental mode, is of much interest. It is sufficient in such cases to recognise the possibility of higher harmonics, without calculating their frequencies or amplitudes in detail.

8.5 Exact solutions of diffusion equations by separation

It is possible, in principle, to solve conduction/diffusion equations in the same way as wave equations, by separation to obtain an eigenvalue equation, which can be solved in terms of eigenfunctions and eigenvalues.

As an example of a conduction problem, consider a large slab of material, thickness L, which is initially at uniform temperature θ_0. At time $t=0$ it is plunged into a cooling liquid at temperature $\theta=0$ (figure 8.3(a)).

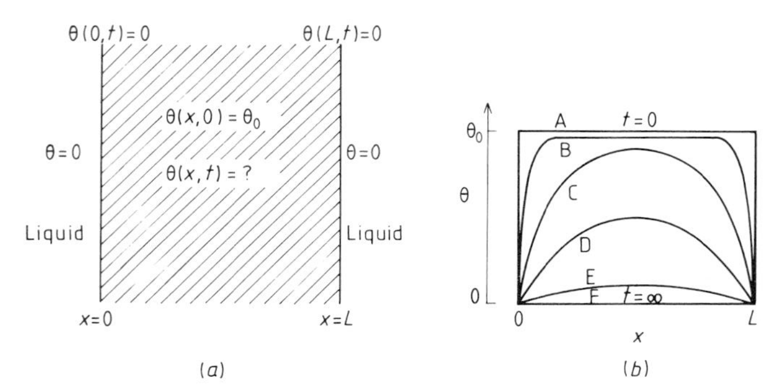

Figure 8.3 (a) Conditions of a cooling slab. Here θ is the excess temperature, initially θ_0 throughout the thickness of the slab from $x=0$ to $x=L$. The exposed surfaces are cooled by a bath at $\theta=0$. (b) Qualitative sketches of the decay of the temperature from its initial profile at $t=0$ (curve A) to its final state $\theta=0$ (all x) at $t=\infty$ (curve F).

The heat entering a small volume element δV can be shown to be proportional to $\partial^2\theta/\partial x^2$, and the subsequent temperature is governed by the heat balance equation

$$\underset{\left[\text{rate of increase of temperature in } \delta V\right.}{\frac{\partial \theta}{\partial t}} \quad \underset{\propto}{=} \quad \underset{\left.\text{heat entering } \delta V\right]}{D\frac{\partial^2 \theta}{\partial x^2}} \tag{8.15}$$

where D is the thermal diffusivity. The boundary conditions are that $\theta = 0$ at $x = 0$ and $x = L$ for all times at and after $t = 0$, so $\theta(0, t) = 0$ and $\theta(L, t) = 0$. There is also an initial condition, that $\theta = \theta_0$ for all x when $t = 0$, so $\theta(x, 0) = \theta_0$.

Figure 8.3(*b*) is a qualitative sketch showing how the temperature profile across the slab would be expected to change with time. For all values of x, the temperature starts from θ_0 and decays to zero as $t \to \infty$, the decay being slowest in the centre. By analogy with linear ODE 1, this smooth decay in time can be suspected to be exponential, so it is plausible to try to separate equation (8.15) by the substitution

$$\theta = X \, \mathrm{e}^{-t/\tau} \tag{8.16}$$

where X is a function of x only.

Substituting this into equation (8.15) and rearranging gives the eigenvalue equation

$$\frac{\mathrm{d}^2 X}{\mathrm{d}x^2} + \left(\frac{1}{D\tau}\right) X = 0 \tag{8.17}$$

together with the boundary conditions

$$X(0) = 0 \qquad X(L) = 0.$$

The eigenfunctions for X are again the set of sine waves satisfying the boundary conditions, and the spatial eigenfunctions are the same as in the previous example, that is,

$$X_n = A_n \sin\left(\frac{n\pi x}{L}\right) \qquad n = 1, 2, \ldots. \tag{8.18}$$

The corresponding eigenvalues of τ can be determined by substituting these spatial eigenfunctions into the eigenvalue equation (8.17). This gives

$$\frac{-A_n n^2 \pi^2}{L^2} \sin\left(\frac{n\pi x}{L}\right) + \left(\frac{1}{D\tau}\right) A_n \sin\left(\frac{n\pi x}{L}\right) = 0.$$

Dividing by $\sin(n\pi x/L)$ gives an equation for the eigenvalues of τ, i.e.

$$\tau_n = \frac{L^2}{n^2 \pi^2 D} \qquad n = 1, 2, \ldots. \tag{8.19}$$

Substituting τ_n and X_n into the trial function (8.16) gives the set of eigenfunctions of the original PDE:

$$\theta_n = A_n \sin\left(\frac{n\pi x}{L}\right) \mathrm{e}^{-t/\tau_n} \tag{8.20}$$

where the decay time constant τ_n of the nth eigenfunction is given by the result (8.19).

8.6 Decay modes

It is possible to interpret the eigenfunctions (8.20) as representing *normal modes of decay*, by analogy with the normal modes of vibration for the wave equation. When t is put equal to 0, in the eigenfunction (8.20),

$$\theta_n(x, 0) = X_n$$

so once again spatial eigenvalues X_n describe the initial shape of the temperature profile needed to excite a normal mode.

To match the initial condition $\theta(x, 0) = \theta_0$ for the problem of §8.5, it is necessary to superpose these normal decay modes. The required solution can be shown to be

$$\theta = \frac{4\theta_0}{\pi}\left[\sin\left(\frac{\pi x}{L}\right)\mathrm{e}^{-t/\tau} + \tfrac{1}{3}\sin\left(\frac{3\pi x}{L}\right)\mathrm{e}^{-9t/\tau} + \tfrac{1}{5}\sin\left(\frac{5\pi x}{L}\right)\mathrm{e}^{-25t/\tau} + \cdots\right] \tag{8.21}$$

where

$$\tau = L^2/\pi^2 D$$

is the time constant for the decay of the lowest mode.

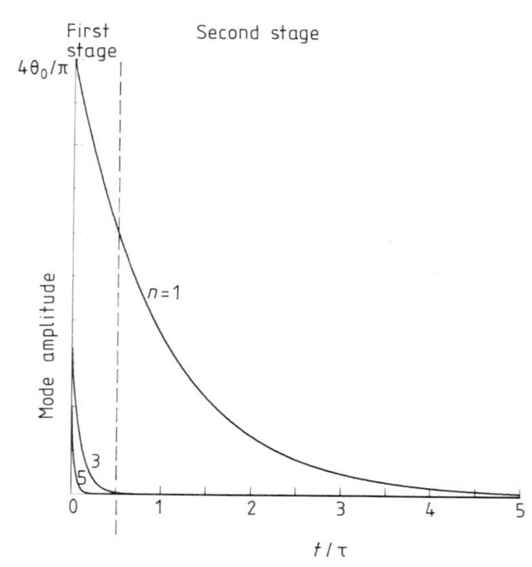

Figure 8.4 Decay of amplitude of the first three modes of the cooling slab shown in figure 8.3 (see function (8.21)). For the fundamental mode ($n = 1$), the amplitude is proportional to $\mathrm{e}^{-t/\tau}$. For the higher modes ($n = 2, 3$), the amplitudes are proportional to $\frac{1}{3}\mathrm{e}^{-9t/\tau}$ and $\frac{1}{5}\mathrm{e}^{-25t/\tau}$ respectively.

The behaviour of the decay modes can be seen in figure 8.4, which shows how the amplitudes of the first three terms of solution (8.21) decay with time. These curves show the rapid collapse of the higher modes during the initial stage of the process, lasting from $t=0$ to $t\sim\frac{1}{2}\tau$. For times greater than this, only the fundamental mode ($n=1$) remains to any significant extent. Physically, this first stage corresponds to the collapse of the outer shoulders of the temperature profile in figure 8.3(b), which changes from curve A through curve B to curve C.

The fundamental decay term is

$$A_1 \sin\left(\frac{\pi x}{L}\right) \mathrm{e}^{-t/\tau}$$

whose shape (a half sine wave) is approximately parabolic. The second stage of decay, from $t\sim\frac{1}{2}\tau$ to $t=\infty$, is dominated by the relatively slow decay of the fundamental mode shown in figure 8.4. Physically, this corresponds to the collapse of the parabola-like profile from curve C in figure 8.3(b) through curves D and E, ultimately reaching $\theta=0$ for all x (curve F).

This example illustrates an important general feature of the solution curves of conduction and diffusion equations. Although the exact solution (8.21) is a series of oscillatory spatial functions, there is no trace of oscillation in the temperature profiles (figure 8.3(b)). Unlike the higher modes of a vibrating string, it is not possible to attach any clear physical meaning to the individual higher modes of this conduction problem; simple conduction/diffusion problems have solutions that are nonoscillatory in space and it is both artificial and physically distasteful to describe them in terms of spatially oscillatory functions. The attractive feature of the QSTF method for such problems is that it uses only nonoscillatory functions, such as exponentials and parabolas, to represent these nonoscillatory solution curves. Examples of the method are given in Chapter 10.

Chapter 9

Estimation of lowest eigenvalues by parabolic trial functions

9.1 Lowest-eigenvalue problems

In this chapter, the QSTF method for estimating the lowest eigenvalue will be introduced, using the parabolic TFs studied in Chapter 5.

The mathematical and physical features that distinguish eigenvalue problems have already been outlined in Chapter 8. Lowest-eigenvalue problems frequently occur in various branches of experimental science, notably those modelled by wave equations. Usually the wave equation is linear, and it can then be separated by the substitution

$$u = X \cos \omega t$$

as described in §8.2. The equation for X is then an eigenvalue equation, and the lowest eigenvalue corresponds to the frequency of the fundamental mode of the oscillations described by the original wave equation. This fundamental frequency is the most significant physical characteristic of an oscillating system, and it is the most easily observed and measured experimentally. In many problems, it is the only information that is required from the model. Even if more detailed information is required, the understanding gained from a TF estimation of the lowest eigenvalue is very useful.

The lowest-eigenvalue problem can also arise more directly, since the mathematical models of some systems turn out to be inherently oscillatory ODE 2 whose solutions must be fitted to boundary conditions in space. A mathematical complication of many eigenvalue equations is the presence of *singular terms* (§9.5). The connection between these terms and boundary conditions at the origin is discussed in physical terms in §9.6.

For reasons that will be explained in §9.3, it is not generally feasible to extend the TF method to higher eigenvalues. This limits the value of the method in some theoretical work, notably in quantum physics. In examples based on experimental systems, however, the restriction to the lowest mode is usually of little consequence, and is compensated by the variety of equations, some quite formidable, that can be tackled to obtain useful results, together with the understanding that the QSTF method fosters. For instance, the final problem treated by the QSTF method in this chapter (§§9.5–9.8) involves an eigenvalue equation that is both singular and nonlinear, a combination of mathematical features, each one of which is often an obstacle to understanding and a source of difficulty.

9.2 Estimation of the lowest eigenvalue of a vibrating string

Before applying the method to more complicated examples, we shall illustrate its use by estimating the lowest eigenvalue for the vibrating string. The simple wave equation that governs its behaviour, i.e.

$$\frac{\partial^2 u}{\partial t^2}=c^2\frac{\partial^2 u}{\partial x^2}$$

has already been solved exactly in Chapter 8. The reader is asked to ignore this exact solution, and concentrate upon the steps of the QSTF method.

The first step for a wave equation is to separate it by the substitution

$$u=X\cos\omega t.$$

For the present example, it was shown in §8.2 that this leads to the eigenvalue equation

$$\frac{\mathrm{d}^2 X}{\mathrm{d}x^2}+\frac{\omega^2}{c^2}X=0 \tag{9.1}$$

with the boundary conditions that the displacement is zero at both ends of the string.

9.2.1 Qualitative sketch

The simplest curve between the two ends that satisfies the boundary conditions is the parabola-like figure shown in figure 9.1. The curve must be symmetrical about the centreline $x=0$, since there is no physical distinction between the two ends of the string, one at $x=-l$ and the other at $x=+l$. For reasons to be discussed in §9.3, we shall assume this curve is the qualitative sketch of the lowest eigenfunction.

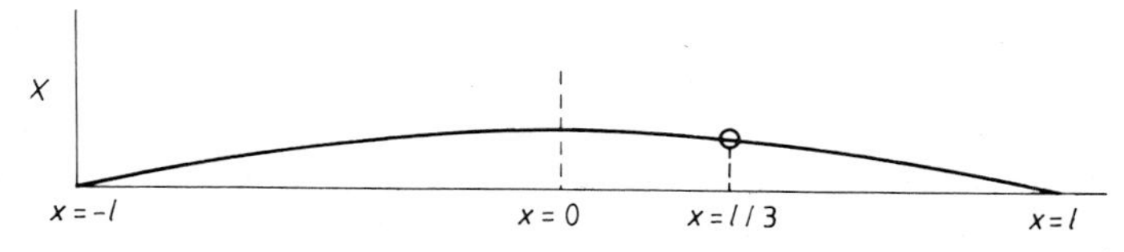

Figure 9.1 Qualitative sketch for the lowest eigenfunction of a vibrating string, also showing the collocation point (small open circle). The ends are chosen at $x=\pm l$ to emphasise the symmetry.

9.2.2 Choice of trial function

Comparison of the sketch with the parabolic functions in §5.1 shows that the TF for the lowest eigenfunction is

$$X^* = A\left(1 - \frac{x^2}{l^2}\right) \tag{9.2}$$

which satisfies the boundary conditions

$$X(-l) = 0 \qquad X(+l) = 0.$$

9.2.3 Substitution of the trial function to form the residual equation

Substituting the TF (9.2) into the eigenvalue equation (9.1) gives the residual equation

$$\frac{-2A}{l^2} + \frac{\omega^2}{c^2} A\left(1 - \frac{x^2}{l^2}\right) = \mathscr{R}. \tag{9.3}$$

If the TF were an exact eigenfunction, the residual could be made zero for all values of x, provided ω was equal to the exact eigenvalue.

Quite generally, substitution of the exact eigenfunction into the equation will always lead to an algebraic equation giving the exact value of the corresponding eigenvalue. It is therefore reasonable to suppose that substitution of an approximate eigenfunction will lead to an approximate value for the corresponding eigenvalue. For approximate TFS, the residual equation cannot be balanced for all x, and therefore must be approximately balanced by collocation, as in Chapter 5. For parabolic TFS in one dimension, the collocation point is one-third of the way across, which is (figure 9.1) at $x = \frac{1}{3}l$. Setting $\mathscr{R} = 0$ and $x = \frac{1}{3}l$ in equation (9.3) gives

$$\frac{-2A}{l^2} + \frac{\omega^2}{c^2} A\left(1 - \frac{(\frac{1}{3}l)^2}{l^2}\right) = 0$$

that is

$$\omega = 1.5c/l = 3.0c/L \tag{9.4}$$

where $L = 2l$ is the length of the string. This estimate agrees with the functionality of the exact result, $\omega = \pi c/L$, and the numerical error is less than 5%.

It will be observed that the QSTF method for finding the eigenvalue follows the same steps—sketch, choose TF, substitute, collocate and solve—that have been used in earlier chapters. The resulting design formula is a relation between the eigenvalue and other parameters of the system. In many cases, the goal is an explicit formula for the eigenvalue; in other cases, the goal may be to find the value of one of the other parameters to achieve a given eigenvalue. For instance, an alternative goal for the vibrating string problem might be to find the length L to achieve a given frequency $f=\omega/2\pi$. The relation (9.4) must then be rearranged to give the design formula

$$L=0.48c/f.$$

9.3 Parabolic trial functions as approximate eigenfunctions

The first step of the TF approximation is to choose a trial function that is a good approximation to the exact eigenfunction for the lowest mode. Before solving any further examples, therefore, we shall consider how this choice can be made, and show why the parabolic TF is suitable in many practical circumstances.

The TFS are chosen, as usual, by making a qualitative sketch and then matching this to the curve of a familiar function. The difficulty is that qualitative sketches cannot always be made by physical intuition, since the concept of eigenfunctions (and still more the shapes of their curves) is quite remote from direct physical experience. For sketching eigenfunctions, therefore, appeal is made to insight gained from simple exact solutions like those discussed in Chapter 8.

Spatial eigenfunctions arise as solutions of spatial oscillatory equations, so they must be smoothly curving waves. These waves must then be fitted between the boundaries, and they must satisfy the boundary conditions. The graphs of the eigenfunctions of any spatial oscillatory equation, however complicated, must therefore look like a set of standing waves. The eigenfunction of the lowest mode is by definition the simplest of this set of curves; it will be the one with the least number of extrema, and also it will cut the zero line the least number of times. In figure 9.2, for example, the curve for $n=1$ has one maximum and no zero between the boundaries, whereas the curve for $n=2$ has two extrema and an intermediate zero; the curves for higher n, such as $n=3$, are even more complicated.

The inference is that the qualitative sketch of the lowest eigenfunction can be made by drawing the simplest curved figure between the boundaries that is consistent with the boundary conditions. (The straight line is excluded, since the eigenfunctions must be wave-like.)

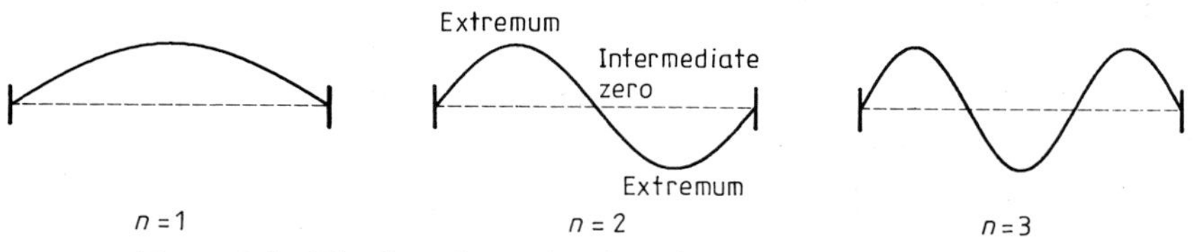

Figure 9.2 The first three eigenfunctions of a vibrating string.

Figure 9.3 shows the application of this rule to two very common boundary conditions. The variety of possible curves indicates that, although the exact shape is not known, it is clear that the boundary conditions will constrain the curves so that it is unlikely that they lie very far from the parabolic curves shown by the full curve. In many problems, therefore, the TF for the lowest eigenfunction will be one of the two parabolic functions described in §5.1. In the present chapter, all the examples have the right boundary conditions for the parabolic TFS to be useful†.

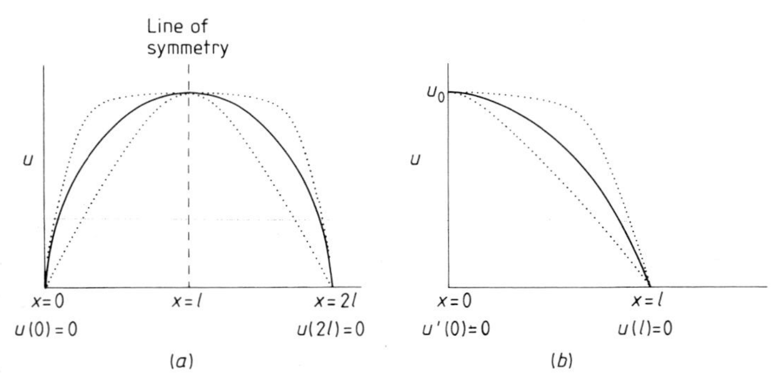

Figure 9.3 Possible curves (dotted) of lowest eigenfunctions that match two common pairs of boundary conditions. The full curves are parabolas.

Understanding the eigenfunctions as a set of standing waves also shows why the QSTF method becomes too complicated when applied to the higher modes. Figure 9.4 shows some possible waves fitted between the boundary conditions $u'(0)=0$, $u(l)=0$ for the second mode, which must be the next simplest curve (with one intermediate zero) after the simple curve already used for the fundamental mode. Without knowing the exact solution, it is impossible to choose between the various shapes, since the position of the zero and the heights and depths of the extrema are all unknown. Corresponding to each of these unknowns, the TF must therefore contain several unknown constants, whose values could be found only by collocating at a corresponding number of points. The simplicity of the single-point collocation is therefore lost, and the labour of the method is not generally justified by the results.

One other aspect of the use of parabolic TFS as lowest eigenfunctions is the choice of collocation point in cylindrical or spherical coordinates. It will be recalled that in one dimension the recommended collocation point is one-third of the way across the x-domain (§5.3). In cylindrical symmetry, the single variable r replaces the two variables x and y. If collocation is to be made one-

† Cosine TFS can also be used (see problem 9.6), but the parabolic TF usually leads to simpler working.

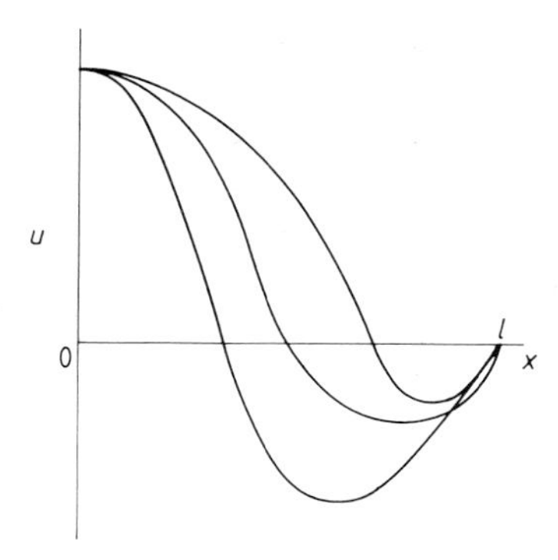

Figure 9.4 Possible curves of second eigenfunctions for the boundary conditions $u'(0)=0$ and $u(l)=0$.

third of the way across each of the x and y domains, that is at $x=R/3$ and $y=R/3$ (where R is the radius of the boundary), then the corresponding collocation point for r is at $r=(x^2+y^2)^{1/2}=0.48R$, or very nearly at $r=\frac{1}{2}R$. A similar argument applies to spherical symmetry, where the corresponding factor is $r=0.58R$. In order to avoid confusion by having too many different rules for collocation, the rule for both cylindrical and spherical symmetries will be taken to be:

> 'Collocate parabolic TFs in cylindrical
> or spherical symmetry at $r=\frac{1}{2}R$.'

This rule will be found to give useful numerical accuracy whenever a parabolic TF is a reasonable match to the qualitative sketches. For cartesian coordinates, the rule: 'collocate one-third of the way across' still applies.

9.4 The tidal periods in lakes

Measurements of water level at the ends of long lakes show that the level rises and falls with a definite period, in the manner of tides on the seashore. The water in lakes is therefore oscillating back and forth, and it is of practical interest to find a design formula for the period of the oscillation.

An equation for the velocity of water in a long lake, obtained from the basic equations of hydrodynamics, is

$$gh\,\frac{\partial}{\partial x}\left(\frac{1}{b}\frac{\partial}{\partial x}(bu)\right)=\frac{\partial^2 u}{\partial t^2} \qquad (9.5)$$

$$\left[\begin{matrix}\text{change in force on}\\ \text{unit mass per second}\end{matrix}=\begin{matrix}\text{change in rate of change of}\\ \text{momentum of unit mass per second}\end{matrix}\right]$$

where u is the velocity in the x direction, parallel to the long axis of the lake, at time t. Here b is the breadth of the lake, which may vary with x; h is its depth, here assumed to be constant; and g is the acceleration of gravity. The centre of the lake is taken as $x=0$ and the two ends are at $x=-l$ and $x=+l$, where the velocity must be zero, since no water flows across the boundaries. Equation (9.5) is identified as a wave equation by the partial differential $\partial^2/\partial t^2$. Since the water is known to move in a periodic fashion, and the goal is to find the period, the equation is separated by the substitution

$$u=X \cos \omega t$$

where X is a function of x only, corresponding to the initial velocity distribution at $t=0$. The substitution gives

$$gh \cos \omega t \frac{\mathrm{d}}{\mathrm{d}x}\left(\frac{1}{b}\frac{\mathrm{d}}{\mathrm{d}x}(bX)\right)=-\omega^2 X \cos \omega t.$$

Dividing by $gh \cos \omega t$ gives the eigenvalue equation

$$\frac{\mathrm{d}}{\mathrm{d}x}\left(\frac{1}{b}\frac{\mathrm{d}}{\mathrm{d}x}(bX)\right)+\left(\frac{\omega^2}{gh}\right)X=0. \tag{9.6}$$

The goal is to find the lowest eigenvalue ω which corresponds to the fundamental period of the oscillation.

The breadth b of the lake can be any function of x. Since most real lakes are of irregular shape (figure 9.5(a)), $b(x)$ can only be represented strictly by a complicated function, or modelled numerically. For high accuracy, therefore, the equation would have to be solved numerically. For approximate purposes, many long lakes can be regarded as either rectangular or elliptical in shape, as shown in figure 9.5(b) and (c).

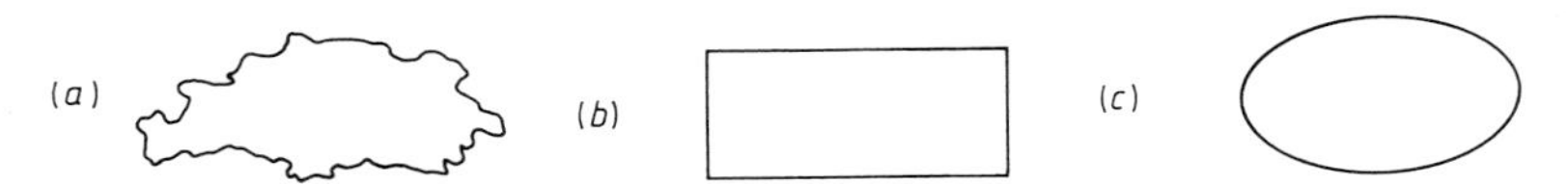

Figure 9.5 Real and idealised plans of lakes: (a) real, (b) rectangular and (c) elliptical.

For the rectangular plan, b is constant, and equation (9.6) reduces to

$$\frac{\mathrm{d}^2 X}{\mathrm{d}x^2}+\left(\frac{\omega^2}{gh}\right)X=0$$

which is identical in form to the eigenvalue equation (8.6) of a vibrating string, except that c^2 is replaced by gh. The boundary conditions ($X=0$ at both ends) are also identical, so the exact lowest eigenvalue of ω is obtained directly from solution (8.9) by putting $n=1$ and $c=(gh)^{1/2}$. This gives

$$\omega = \frac{\pi(gh)^{1/2}}{L} \tag{9.7}$$

for the lowest eigenvalue of ω for the rectangular lake of overall length L.

For shapes other than rectangular, the breadth varies with x, and complicated equations that have no obvious solution are obtained. For an elliptical lake, for instance, b varies from a maximum breadth b_0 at the centre ($x=0$) to zero at each end ($x=\pm l$), and we can write

$$b = b_0\left(1-\frac{x^2}{l^2}\right)^{1/2}. \tag{9.8}$$

The equation to be solved is then

$$\frac{\mathrm{d}}{\mathrm{d}x}\left\{\frac{1}{(1-x^2/l^2)^{1/2}}\frac{\mathrm{d}}{\mathrm{d}x}\left[\left(1-\frac{x^2}{l^2}\right)^{1/2} X\right]\right\}+\left(\frac{\omega^2}{gh}\right)X=0. \tag{9.9}$$

The boundary conditions upon X are $X(-l)=0$ and $X(+l)=0$.

The TF solution for the lowest eigenvalue of ω for this more complicated equation follows exactly the steps used for the vibrating string example.

9.4.1 Solution

The simplest curve that can be drawn which agrees with $X=0$ at $x=-l$ and $x=+l$ is once again approximately a parabola (figure 9.6). This will be taken as the qualitative sketch of the lowest eigenfunction.

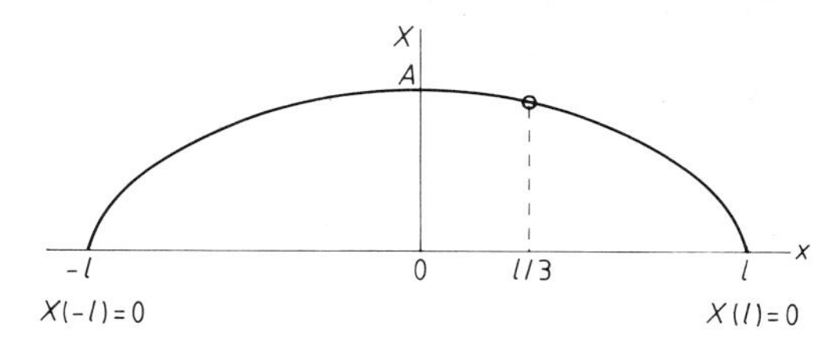

Figure 9.6 Qualitative sketch for the lowest spatial eigenfunction of tidal motion in a lake, also showing the collocation point.

Comparison with figure 5.1 suggests the corresponding TF

$$X^* = A\left(1-\frac{x^2}{l^2}\right).$$

Substituting this TF into equation (9.6) gives the residual equation

$$\frac{\mathrm{d}}{\mathrm{d}x}\left\{\frac{1}{b}\frac{\mathrm{d}}{\mathrm{d}x}\left[bA\left(1-\frac{x^2}{l^2}\right)\right]\right\}+\frac{\omega^2}{gh}A\left(1-\frac{x^2}{l^2}\right)=\mathscr{R}.$$

With b given by equation (9.8), this simplifies to

$$\frac{-3A}{l^2}+\frac{\omega^2 A}{gh}\left(1-\frac{x^2}{l^2}\right)=\mathscr{R}. \tag{9.10}$$

Since this is in cartesian coordinates, it is collocated as usual at one-third of the way across, that is at $x=\frac{1}{3}l$ (see figure 9.6). The result is

$$\frac{-3}{l^2}+\frac{8}{9}\frac{\omega^2}{gh}=0$$

whence the design formula for the lowest eigenvalue of ω is

$$\omega=\frac{1.84(gh)^{1/2}}{l}=\frac{3.7(gh)^{1/2}}{L} \tag{9.11}$$

where $L=2l$ is the length of the lake. The design formula for the tidal period of an elliptical lake is therefore

$$T=\frac{1.7L}{(gh)^{1/2}}. \tag{9.12}$$

The design formula for ω for the elliptical lake does not contain b_0, the maximum breadth of the lake. The periods of a circular lake and one with a very thin elliptical plan are therefore, perhaps rather surprisingly, exactly the same provided both have the same 'length' L. However, comparison of the formulae (9.11) and (9.7) for elliptical and rectangular lakes shows that the numerical factors are in the ratio 3.7/3.14, a difference of some 18%. This suggests that the calculated period may be sensitive to the approximating shape used to match the irregular plan view of natural lakes. This example illustrates one of the uses of the TF method mentioned in the Introduction (§1.5); it supplies preliminary understanding before attempting a computed numerical solution. It shows that computer models of lakes will achieve numerical results no better than the TF method unless the plan is modelled with high precision, rather than simplifying the plan to some convenient geometric shape.

9.5 The fluorescent tube: mathematical model

An example of an eigenvalue equation which arises directly, rather than by separation of a wave equation, originates in the mathematical model of the electron density in a fluorescent tube. Besides illustrating this point, the physical equation for the model introduces the important subject of singularities, which will be discussed in more detail in the next section.

Fluorescent lamps are essentially discharge tubes. Most of the length of the tube is occupied by a positive column discharge, in which electrons are generated by ionisation, diffuse radially to the wall of the tube and are lost there by recombination with positive ions.

In the positive column, the electron density n depends only upon the radial distance r from the axis, since the discharge is axisymmetric (no ϕ dependence) and independent of position along the tube, except very near the ends (no z dependence). It is therefore convenient to work in cylindrical coordinates (appendix 5). The governing equation for the balance of losses and gains of electrons in the small volume element δv is

$$\frac{\partial n}{\partial t} = \nu_i n + D\left(\frac{\partial^2 n}{\partial r^2} + \frac{1}{r}\frac{\partial n}{\partial r}\right). \tag{9.13}$$

[rate of gain of electrons in δv = rate of generation by ionisation in δv − rate of loss from δv by diffusion]

The assumption is that new free electrons are generated by electron impact, and the generation rate is therefore proportional to the electron density. Here ν_i is the coefficient of proportionality for ionisation, and D is the diffusion constant of electrons in the gas mixture.

If the discharge is stable, n is constant with time, so the governing equation for a stable discharge is obtained by putting $\partial n/\partial t = 0$ in equation (9.13). This gives the ODE

$$\frac{d^2 n}{dr^2} + \frac{1}{r}\frac{dn}{dr} + \left(\frac{\nu_i}{D}\right) n = 0. \tag{9.14}$$

The positive sign of the restoring term coefficient shows that the equation is inherently oscillatory. The solutions are standing waves which must be fitted between the boundary conditions; they are therefore eigenfunctions and the equation is an eigenvalue equation. The goal is to obtain a design formula for the lowest eigenvalue of the coefficient ν_i; this will be shown in §9.7 to be directly related to the voltage required to maintain the stable discharge.

The reader may recognise the governing equation as a *zero-order Bessel equation*, and the corresponding eigenfunctions as zero-order Bessel functions. He is asked to ignore this exact solution for the time being; the advantage of the QSTF method will be apparent in the next example, when a nonlinear term is added and solution in terms of tabulated functions is no longer possible.

Before an attempt is made to solve equation (9.14), it is important to recognise two new features illustrated by this problem. First, the equation is an example of an eigenvalue equation which has arisen directly, rather than by the separation of a wave equation. Secondly, the equation contains the term

$$\frac{1}{r}\frac{dn}{dr}$$

which has a negative power, r^{-1}, of the independent variable. When $r=0$, the term becomes

$$\left(\frac{\mathrm{d}n}{\mathrm{d}r}\right)\bigg/0$$

which is infinitely large unless $\mathrm{d}n/\mathrm{d}r$ is also zero, in which case the term has the indeterminate value 0/0. Terms that become infinite or indeterminate at some point in the domain of the independent variable are called *singular terms*, and the equation is said to have a *singularity* at that point.

Physical equations like equation (9.14) which have singular terms need special treatment. The next section considers some of the peculiarities of singular equations to a depth sufficient to allow problems involving singularities to be solved by the QSTF method.

9.6 Singularities at fictitious boundaries

In the fluorescent lamp described in the previous section, the only physical boundary confining the positive column is the inner surface of the cylindrical tube. Physically, therefore, there is only one boundary, the tube wall, which is located at $r=R$ (figure 9.7). Since the concentration of electrons at the wall is zero, there is one explicit physical boundary condition.

$$n(R)=0.$$

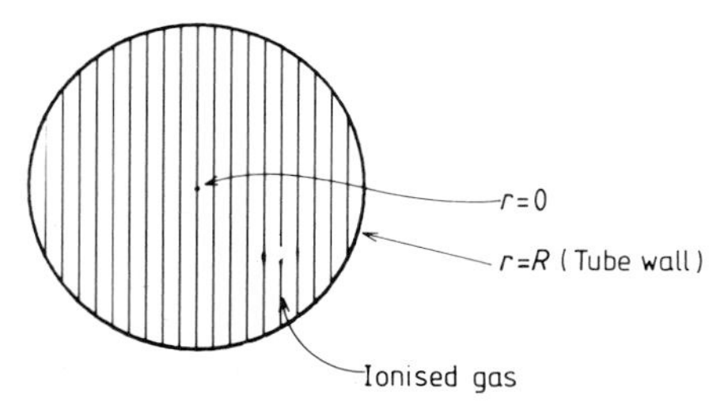

Figure 9.7 Cross-section of a fluorescent tube, showing the real physical boundary at $r=R$ and the fictitious mathematical boundary at $r=0$.

Mathematically, however, the domain of r has two boundaries, for since r cannot be negative it ranges from a mathematical boundary at $r=0$ to one at $r=R$. The mathematical boundary at $r=R$ corresponds to the real physical boundary at the tube wall, where there is an abrupt physical change from gas to solid. By contrast, there is no physical object at the axis to disturb the continuity of the gaseous phase and present a real physical boundary near $r=0$; the mathematical boundary at $r=0$ is therefore fictitious. It is a consequence of the use of cylindrical coordinates as a convenient way of formulating the mathematical model.

The point $r=0$, however, is not only a fictitious boundary, but is also a singular point, as discussed at the end of the last section. This combination of circumstances, in which a mathematical boundary has no obvious physical boundary condition, and is also a singular point, occurs frequently in ODE 2, particularly when the model uses cylindrical or spherical coordinates. To solve the present equation, some method of finding this missing second boundary condition must be devised, since the standing waves corresponding to the eigenfunctions have to be fitted to two boundary conditions.

We are accustomed to accept that measurable physical properties, such as temperature or concentration, are always finite, since we cannot conceive of any effects that will make them infinite. Further reflection will show that there are likewise no physical effects strong enough to produce infinite rates of change in a physical quantity, either with space or time. In other words, the first differential of any physical quantity is finite. Similar reasoning shows that the second differential must also be finite, since the rate of change of a gradient cannot be infinite. Generally it is an excellent hypothesis that in any equation, if the dependent variable is a measurable quantity having physical dimensions, then it and all its differentials with respect to the independent variables are continuous and finite. Investigation of apparent exceptions to this rule will usually reveal either that the dependent variable is dimensionless (for instance, it is possible for the dimensionless quantity $\tan\theta$ to become infinite), or that the model has approximated a very strong influence by an infinite one. ('Infinite' forces arise, for instance, in potential theory at 'point sources', which are approximations for small charged objects. Another example occurs in problem 5.5, where a large space charge at the cathode of a diode has been modelled as though it were infinite. As a consequence, there is a singularity at $x=0$, and the equation predicts erroneously that the second differential V'' is infinite at that point. Notice however that in both these cases there is no fictitious boundary at the singular point, since there is a real physical object bounding the domain of the independent variable.)

For physical equations that have only one singular term, giving a singularity at a fictitious boundary, the rule that the *second differential is finite* will reveal the missing boundary condition with surprising ease. To illustrate the rule for the fluorescent tube equation (9.14), first it is written as

$$\frac{\mathrm{d}^2 n}{\mathrm{d}r^2}=-\frac{1}{r}\frac{\mathrm{d}n}{\mathrm{d}r}-\frac{v_{\mathrm{i}}}{D}n.$$

At the singular point, $\mathrm{d}^2n/\mathrm{d}r^2$ must remain finite, so the singular term on the RHS

$$\frac{1}{r}\frac{\mathrm{d}n}{\mathrm{d}r}\rightarrow\left(\frac{\mathrm{d}n}{\mathrm{d}r}\right)\Big/0$$

must not become infinite. The only way in which this infinity can be avoided is if dn/dr also tends to zero as $r \rightarrow 0$; that is to say, if

$$n'(0)=0.$$

This, then, is the condition at the fictitious boundary at $r=0$ which, together with the physical boundary condition $n(R)=0$, provides the two boundary conditions needed to set up the eigenfunctions of a second-order eigenvalue equation.

The argument can be illustrated graphically by considering figure 9.8. This shows three curves that might represent the concentration of electrons across a diameter of a fluorescent tube. All three curves satisfy the boundary condition $n(R)=0$, and possess the expected symmetry about the tube axis. However, figure 9.8(b) and (c) both have an infinitely sharp rate of change of gradient at $r=0$; this corresponds to n'' being infinite, so the rule excludes these curves from consideration as eigenfunctions. This leaves figure 9.8(a), which is consistent with the boundary condition at $r=0$, i.e.

$$n'(0)=0 \tag{9.15}$$

already found above.

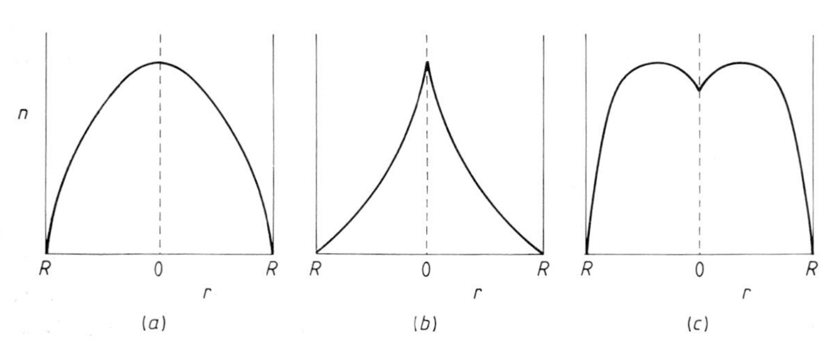

Figure 9.8 Hypothetical curves of electron density along the diameter of a fluorescent tube.

Sometimes an equation contains two singular terms. The simple 'second differential is finite' rule cannot be used in such cases, and the boundary conditions must be found by the more formal time-consuming method of Frobenius. Fortunately, this is not a common circumstance in equations arising directly from mathematical models, and the simple method suggested here works well for all the subsequent examples, even when the equations are nonlinear.

9.7 Maintaining voltage of a fluorescent tube

The eigenvalue equation (9.14) was developed in §9.5 as a simple model for the positive column discharge which occupies almost the entire length of a fluorescent tube. The equation for the electron density n is equation (9.14), i.e.

$$\frac{d^2n}{dr^2}+\frac{1}{r}\frac{dn}{dr}+\frac{v_i}{D}n=0.$$

The boundary conditions deduced in the previous sections are $n(R)=0$ and $n'(0)=0$.

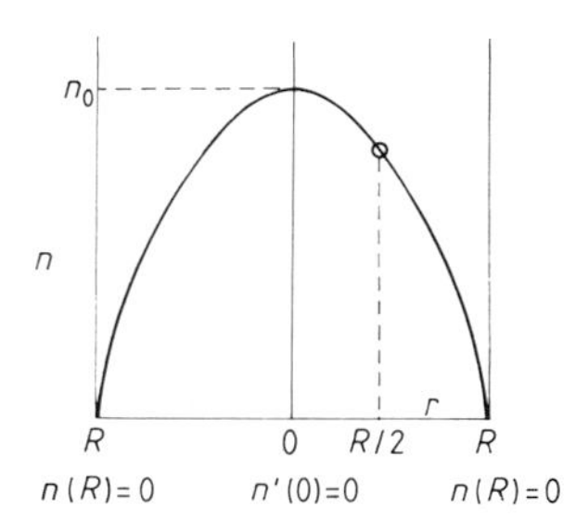

Figure 9.9 Qualitative sketch for the lowest eigenfunction of electron density in a positive column, also showing the collocation point (small open circle).

The simplest curve that can be drawn which agrees with these two boundary conditions is parabola-like (figure 9.9). From §5.1 the TF for the lowest eigenfunction corresponding to this qualitative sketch is

$$n^*=n_0\left(1-\frac{r^2}{R^2}\right) \tag{9.16}$$

where n_0 is the electron concentration at $r=0$. The TF contains the single parameter n_0, and the equation contains the unknown v_i. The goal in the first instance is to find a design formula for the lowest eigenvalue of v_i; this will be shown in subsequent discussion to be directly related to the voltage required to maintain a stable discharge. Substitution of the TF (9.16) into equation (9.14) leads to the residual equation

$$\frac{-2n_0}{R^2}-\frac{2n_0}{R^2}+\left(\frac{v_i}{D}\right)n_0\left(1-\frac{r^2}{R^2}\right)=\mathcal{R}. \tag{9.17}$$

Collocation at $r/R=\frac{1}{2}$ (polar coordinates; see §9.3) gives

$$\frac{-4n_0}{R^2}+\frac{v_i}{D}n_0\left(\frac{3}{4}\right)=0$$

whose solution,

$$v_i=\frac{16}{3}\left(\frac{D}{R^2}\right)=5.3\left(\frac{D}{R^2}\right) \tag{9.18}$$

is the required estimate of the lowest eigenvalue of v_i.

9.7.1 Discussion—the stable maintaining voltage

The eigenvalue of v_i given by formula (9.18) is the lowest of a set of possible eigenvalues. Mathematically, a solution of equation (9.14) is possible if v_i is any member of the set, but in practice the discharge will adapt itself to run in the lowest mode, since this corresponds to the lowest stored energy. The mathematical model of a steady-state positive column therefore has only one practical solution, and this solution is only possible if v_i has the value given by the lowest eigenvalue (9.18). Physically, this means that to be stable the discharge must adapt itself so that v_i has just this value, and this is achieved by controlling the voltage applied to the tube.

As might be expected, the generation constant v_i depends upon the voltage applied to the tube and very roughly obeys the linear relation

$$v_i = AV/pL \tag{9.19}$$

where A is a constant depending on the gas, p is the gas pressure and L is the length of the positive column discharge. Equating the two formulae (9.18) and (9.19) for v_i, and solving for V gives

$$V = 5.3\,\frac{pLD}{AR^2}. \tag{9.20}$$

This is the design formula for the *maintaining voltage* that must be applied to the tube to sustain a stable discharge. In practice, this is achieved by having a resistor in series with the tube. Any tendency for the tube voltage to increase above the maintaining voltage results in an increase in v_i above the eigenvalue for the steady state. The current through the tube would therefore start to increase, which in turn would cause the voltage drop across the external resistor to increase, thereby decreasing the voltage applied to the tube. Provided there is some external resistance, the tube voltage is therefore self-stabilising at the exact value to satisfy the eigenvalue condition.

The constants D and A in the design formula (9.20) depend upon the gas used to fill the tube. For a fixed gas filling and pressure, the design formula shows that the maintaining voltage is related to the tube length L and radius R by

$$V \propto L/R^2.$$

In designing tubes of different length, their radii may be scaled using this formula, so that all may use the same optimum gas filling and still have the same maintaining voltage.

One defect of this linear model is that the design formula suggests that the maintaining voltage is independent of the tube current. Experimentally it is found that the maintaining voltage for a positive column discharge varies slightly as the tube current is altered. In the next section, a nonlinear model of the positive column will be investigated. The effect is analogous to that observed in §6.3, where the addition of a nonlinear term accounted for the observed dependence of oscillator frequency upon amplitude. In this case, the nonlinearity accounts for the *V–i* dependence, and the QSTF method is invaluable in dealing with the extra complications caused by the nonlinear term.

9.8 The nonlinear positive column discharge

The linear equation (9.14) developed in §9.5 assumes that the electrons in a positive column discharge are generated by direct ionisation of gas atoms caused by the impact of fast electrons. However, it is known that secondary ionisation processes involving excited atoms may occur, and these effects, which tend to increase electron generation, depend upon the square of the electron density n. Recombination between electrons and positive ions may also take place, the rate again being proportional to n^2, though in this case the effect is to reduce the electron concentration. Both these additional processes can be included in the model by adding a square-law term to equation (9.14), giving

$$\frac{\mathrm{d}^2 n}{\mathrm{d}r^2}+\frac{1}{r}\frac{\mathrm{d}n}{\mathrm{d}r}+\frac{v_\mathrm{i}}{D}n+\beta n^2=0 \qquad (9.21)$$

where β may be positive or negative depending upon whether the secondary ionisation or the recombination process predominates. The equation, though nonlinear, is still an eigenvalue equation, since so long as β is small it is inherently oscillatory and its solutions are therefore standing waves to be fitted to the boundary conditions. The rule of §9.6 that the second differential is finite can again be used to establish that the concentration gradient $\mathrm{d}n/\mathrm{d}r$ must be zero when $r=0$. It follows that the qualitative sketch is the same as that for the linear case (figure 9.9), and the TF for the lowest eigenfunction is again

$$n^*=n_0\left(1-\frac{r^2}{R^2}\right).$$

9.8.1 Estimation of lowest eigenvalue

Substitution of the TF into equation (9.21) gives the residual equation, which now contains an additional nonlinear term:

$$\frac{-4n_0}{R^2}+\left(\frac{v_\mathrm{i}}{D}\right)n_0\left(1-\frac{r^2}{R^2}\right)+\beta n_0^2\left(1-\frac{r^2}{R^2}\right)^2=\mathscr{R}. \qquad (9.22)$$

Collocation at $r/R=\frac{1}{2}$ gives

$$\frac{-4n_0}{R^2}+\frac{3}{4}\left(\frac{v_\mathrm{i}}{D}\right)n_0+\frac{9}{16}\beta n_0^2=0$$

whence

$$\nu_i = \frac{16}{3}\frac{D}{R^2} - \frac{3}{4}\beta D n_0 \tag{9.23}$$

is the lowest eigenvalue of ν_i. If the secondary effects are unimportant, $\beta = 0$, and the formula for ν_i becomes the same as the result (9.18) already obtained for the linear case.

9.8.2 The maintaining voltage

The relation between the eigenvalue of ν_i and the maintaining voltage is similar to that already discussed in §9.7. The approximate linear relation (9.19)

$$\nu_i = AV/pL$$

can again be used, but a relation is also needed for β. If secondary ionisation is the predominant process (as would be expected in pure inert gases), β increases with voltage; to seek understanding rather than an exact formula, it is again sufficient to assume a linear relation, writing

$$\beta = BV/pL$$

where B is a constant depending on the gas. Substituting these relations for ν_i and β into the result (9.23), and solving for V gives

$$V = \frac{5.3pLD/AR^2}{1 + \frac{3}{4}BDn_0/A}.$$

The electron current density, n_0, at the axis of the tube is proportional to the current passing through the tube, so for a given tube and gas filling the variation of V with current is

$$V = \frac{K_1}{1 + K_2 i} \tag{9.24}$$

where K_1 and K_2 are constants for a given tube. Formula (9.24) predicts that, as the current through the tube increases, the voltage needed to sustain a stable discharge falls; this is to be expected physically, since increasing the current increases generation of electrons by the secondary ionisation process.

The technically important Penning gas mixtures contrast with pure, inert gases in that the main secondary process expected is recombination in the gas rather than secondary ionisation. In this case, β is a negative constant independent of voltage. The maintaining voltage is then found, by substituting formula (9.19) into formula (9.23), to be

$$V = \frac{16}{3}\frac{pLD}{AR^2} - \frac{3}{4}\frac{\beta pLD}{A} n_0.$$

Since n_0 is proportional to i, the maintaining voltage for Penning mixtures is predicted to be

$$V = K_1 + K_3 i \tag{9.25}$$

where K_3 is a positive constant, since β is negative.

The variation of maintaining voltage with current predicted by formula (9.24) for pure inert gases and by (9.25) for Penning mixtures is sketched in figure 9.10, together with the linear theory result (9.20). Experimental curves often agree with the two curve shapes predicted by this approximate nonlinear theory.

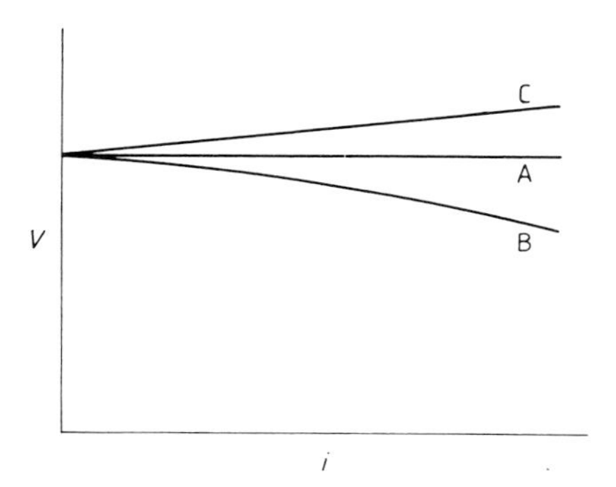

Figure 9.10 Voltage–current characteristics for a fluorescent tube deduced from TF design formulae: curve A, linear model (formula (9.20)); curve B, nonlinear model for pure, inert gas filling (formula (9.24)); curve C, nonlinear model for Penning gas mixture (formula (9.25)).

9.9 Buckling of a vertical column

9.9.1 Statement of the problem

An important eigenvalue problem in structural mechanics concerns the stability of a vertical column or pillar under its own weight. A short, thick column (figure 9.11 (*a*)) is completely stable; a tall, thin column (figure 9.11 (*b*)) is liable to buckle under its own weight since any slight disturbance from the vertical equilibrium position is likely to result in collapse in one or more 'buckling modes' (figure 9.11 (*c*) and (*d*)).

The problem considered here is to estimate the critical condition for buckling of a vertical column of constant cross-section, rigidly built in at its lower end. The buckling modes illustrated in figure 9.11 (*c*) and (*d*) resemble standing waves, and it can be shown that for small displacements they are the

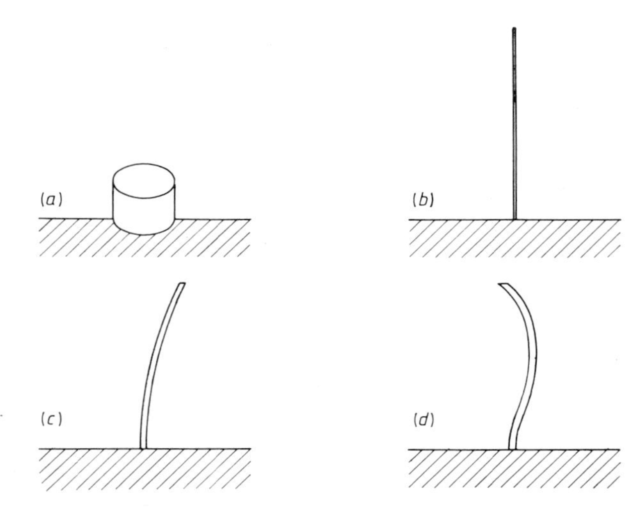

Figure 9.11 (*a*) A short, thick column stable against buckling. (*b*) A tall, thin column liable to buckle. (*c*) Simplest mode of buckling. (*d*) Second mode of buckling.

eigenfunctions of the equation

$$\frac{d^2\alpha}{dx^2}+kx\alpha=0 \tag{9.26}$$

where α is the angle from the vertical and x is the vertical distance measured from the top of the column (figure 9.12). Here

$$k=\mu g/EI \tag{9.27}$$

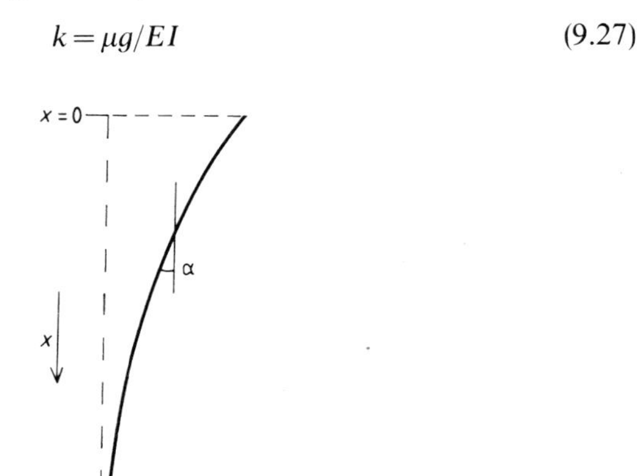

Figure 9.12 Geometry of column buckling in lowest mode.

where μ is the mass per unit length, g is the acceleration due to gravity, E is Young's modulus for the material of the column and I is the cross-sectional moment of inertia.

The boundary conditions are

$$\alpha(L)=0 \qquad \text{(the base of the column is vertical)} \tag{9.28}$$

and

$$\alpha'(0)=0 \qquad \text{(there is nothing to cause bending at the top of the column).} \tag{9.29}$$

For a thick, stiff column, both I and E are large, and so k is small. Physically, the column is stable because the only solution satisfying the equation and boundary conditions is one for which $\alpha=0$ everywhere, i.e. the column is straight and vertical. If I and E are reduced, the value of k increases until a state is reached in which a solution is possible corresponding to a nonzero value of α, i.e. to the column being curved. This is the onset of buckling in the first mode. Finding the critical condition at which buckling first occurs therefore corresponds to finding a design formula for the lowest eigenvalue of equation (9.26) subject to the boundary conditions (9.28) and (9.29).

9.9.2 *Solution by parabolic trial functions*

Figure 9.13 shows the qualitative sketch for the lowest eigenfunction, using the variation of α with x shown in figure 9.12, together with the boundary conditions.

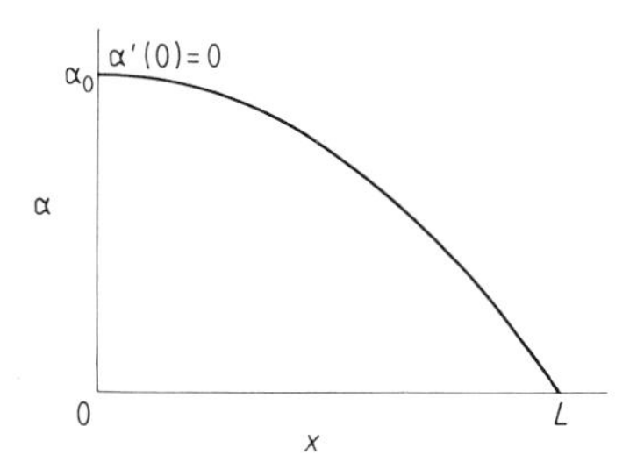

Figure 9.13 Qualitative sketch for the lowest spatial eigenfunction of a buckling column, where α is the angle to the vertical (see figure 9.12).

The shape of the qualitative sketch is approximately parabolic, and this suggests the corresponding standard parabolic TF

$$\alpha^* = \alpha_0\left(1-\frac{x^2}{L^2}\right). \tag{9.30}$$

The residual equation obtained by substituting this TF into equation (9.30) is

$$\frac{-2\alpha_0}{L^2}+kx\alpha_0\left(1-\frac{x^2}{L^2}\right)=\mathscr{R}.$$

Collocation at $x=\frac{1}{3}L$ gives

$$\frac{-2\alpha_0}{L^2}+\frac{kL}{3}\alpha_0(1-\tfrac{1}{9})=0$$

whence the length L and lowest eigenvalue k are predicted to satisfy the relation

$$kL^3=\tfrac{27}{4}=6.75. \tag{9.31}$$

9.9.3 Discussion

The design relation (9.31) can be arranged in two ways: either as a design formula for the lowest eigenvalue of k for a column of given length,

$$k=6.75/L^3 \tag{9.32}$$

or to give the critical length L_b for buckling in the lowest mode for a column for which k is known,

$$L_b=1.9/k^{1/3}. \tag{9.33}$$

The design formulae (9.32) and (9.33) are examples of the alternative goals of an eigenvalue problem mentioned in §9.2. A practical design situation might call for a column of given height, in which case formula (9.32), together with the relation

$$k=\mu g/EI$$

allows the material and cross-section of the column to be chosen. Alternatively, the designer might want to know how high a column can safely be made using specified material of known cross-section. In terms of the original constants, the minimum buckling length given by formula (9.33) is

$$L_b=1.9\left(\frac{EI}{\mu g}\right)^{1/3}.$$

Since buckling first occurs in this mode, a column shorter than this cannot theoretically buckle if it is given a small sideways displacement. In practice, a safety factor of at least 2 is likely to be included in any responsible design, so the discrepancy between the factor of 1.9 in this formula and 1.99 found by exact solution is of little consequence.

Problems

9.1

The wave equation for a string that is heavier at the centre than at the ends is

$$\frac{\partial^2 y}{\partial x^2}=\left(\frac{\mu_0}{T}\right)\left(1-\frac{x^2}{a^2}\right)\frac{\partial^2 y}{\partial t^2}$$

where the mass per unit length is given by $\mu_0(1-x^2/a^2)$. Here a is a measure of the taper, and $a \gg l$, where $2l$ is the length of the string. Distance x is measured from the centre of the string, so the fixed ends are at $x=\pm l$, where the boundary conditions are $y(\pm l)=0$. T is the tension.

(a) Separate the equation. [§8.2] If the eigenvalue equation is written as

$$\frac{\mathrm{d}^2 X}{\mathrm{d}x^2}+k^2\left(1-\frac{x^2}{a^2}\right)X=0$$

what is the value of k in terms of ω, μ_0 and T?

(b) Make a qualitative sketch of the lowest eigenfunction, mark in the collocation points and write down a suitable TF.

(c) Substitute, collocate and hence derive a design formula for the lowest eigenvalue of k.

(d) Use the result for k to obtain the design formula

$$f=f_0\left(1+\frac{l^2}{18a^2}\right)$$

where f_0 is the frequency for a uniform string.

9.2

The governing equation for the deflection y of an axially loaded, tapered column (figure 9.14) is

$$\frac{\mathrm{d}^2 y}{\mathrm{d}x^2}+\frac{ky}{b^2(1-x/b)^4}=0.$$

Always, $x<b$, so there is no singularity between $x=0$ and $x=L$. The boundary conditions are $y'(0)=0$ and $y(L)=0$. (The deflection is measured relative to the free end.)

(a) Write down a suitable TF for the lowest eigenfunction.

(b) Substitute, collocate and find a design formula for the critical value of k for buckling in the lowest mode. [§9.9]

(c) Find the numerical value of k for a 'half-tapered' column, for which $L=b/2$. (The computed result for comparison is 4.10.)

9.3

The wave equation (8.4) for a vibrating string is correct only for vanishingly

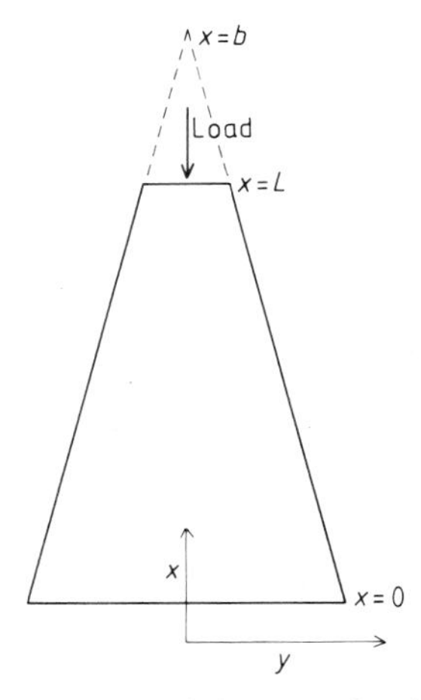

Figure 9.14 Loaded, tapered column.

small displacements. For moderate amplitudes of vibration, a more accurate equation is the nonlinear PDE 2

$$\frac{\partial^2 u}{\partial t^2}=c^2\frac{\partial^2 u}{\partial x^2}\left[1-\frac{3}{2}\left(\frac{\partial u}{\partial x}\right)^2\right].$$

(This is the first example of a nonlinear wave equation. The correction in the bracket is analogous to the nonlinear correction to the pendulum equation discussed in §5.5. It is a common difficulty in solving nonlinear PDEs that they cannot be separated.)

(a) Substitute $u=U(x)\cos\omega t$ to obtain the residual equation

$$\frac{d^2U}{dx^2}\cos\omega t-\frac{3}{2}\left(\frac{dU}{dx}\right)^2\frac{d^2U}{dx^2}\cos^3\omega t=-\frac{\omega^2}{c^2}U\cos\omega t+\mathscr{R}.$$

(Because of the nonlinear term, this does not balance for all values of t as in the case of the linear equation and t cannot be eliminated by simply dividing by $\cos\omega t$. However, harmonic balance (see Chapter 6) can be used to obtain an approximate eigenvalue equation.)

(b) After first substituting

$$\cos^3\omega t=\tfrac{3}{4}\cos\omega t+\tfrac{1}{4}\cos 3\omega t$$

use harmonic balance of the $\cos\omega t$ terms to obtain the equation

$$\left[1-\frac{9}{8}\left(\frac{dU}{dx}\right)^2\right]\frac{d^2U}{dx^2}+\frac{\omega^2}{c^2}U=0.$$

(c) Write this equation in the form (2.25) (with x in place of t), and show that for dU/dx small this is an oscillatory equation. The qualitative sketch is the same as figure 9.1, and the TF is

$$U^* = U_0\left(1-\frac{x^2}{l^2}\right)$$

with the collocation point at $x/l = \frac{1}{3}$ (cartesian coordinates).

(d) Substitute, collocate and show that the design formula for the lowest eigenvalue, assuming U_0/l small, is

$$\omega = \frac{3}{2}\frac{c}{l}\left(1-\frac{1}{4}\frac{U_0^2}{l^2}\right)$$

or with $2l = L$

$$\omega = \frac{3c}{L}\left[1-\left(\frac{U_0}{L}\right)^2\right].$$

(U_0/L is analogous to θ_0 in the result (6.11), but the correction for amplitude for the vibrating string has a much larger coefficient than for the pendulum. When $U_0/L = \frac{1}{4}$ the shift in frequency is one part in 16, which corresponds to a very perceptible fall in pitch of about a semitone from the pitch for $U_0/L \approx 0$.)

9.4

If the effect of the drag of the air on a vibrating string is included, the governing equation (8.4) is modified to

$$\frac{\partial^2 u}{\partial t^2} = c^2\frac{\partial^2 u}{\partial x^2} - b\left|\frac{\partial u}{\partial t}\right|\frac{\partial u}{\partial t}.$$

[acceleration $\propto$ restoring force $-$ drag force]

The drag force causes damping of the motion of the string, and is assumed to obey the same quadratic law as that used for the air-damped pendulum in problem 6.7. The equation is nonlinear and will not separate in the classical fashion. Following the same argument as that used in problem 6.6, take the TF as

$$u^* = U(x)(1-kt/T)\cos\omega t$$

where T is the period and k is the fractional loss of amplitude in one period.

(a) Find $\partial u^*/\partial t$.

(b) Show that for b and k small (so that products bk can be neglected),

$$b\left|\frac{\partial u^*}{\partial t}\right|\frac{\partial u^*}{\partial t} = -bU^2\omega^2|\sin\omega t|\sin\omega t$$

$$= -bU^2\omega^2\left(\frac{8}{3\pi}\right)\sin\omega t + \text{(higher harmonics)}.$$

[Appendix 6]

(c) Hence show that the residual equation, ignoring terms in multiple angles, is

$$U\left[\frac{2k\omega}{T}\sin\omega t-\omega^2\left(1-k\frac{t}{T}\right)\cos\omega t\right]$$
$$=c^2\frac{\mathrm{d}^2U}{\mathrm{d}x^2}\left(1-k\frac{t}{T}\right)\cos\omega t+bU^2\omega^2\left(\frac{8}{3\pi}\right)\sin\omega t+\mathscr{R}.$$

(d) The residual is clearly an oscillating function of t, and also a function of x. Write it as

$$\mathscr{R}=\mathscr{R}_\mathrm{s}(x)\sin\omega t+\mathscr{R}_\mathrm{c}(x)\cos\omega t$$

and hence show that harmonic balance of the sine term gives

$$\frac{2k\omega}{T}=bU^2\omega^2\frac{8}{3\pi}+\mathscr{R}_\mathrm{s}(x).$$

(e) The qualitative sketch of U/x and collocation point are the same as in figure 9.1, and the TF is

$$U^*=A_0\left(1-\frac{x^2}{l^2}\right).$$

Substitute, collocate (noting that the period $T=2\pi/\omega$) and show that the design formula for the fractional amplitude lost in a period is

$$\delta A/A_0=k=\tfrac{64}{27}bA_0.$$

9.5

The governing equation for the deflection u of the skin of an airtight kettledrum is

$$\frac{\partial^2u}{\partial t^2}\quad=\quad c^2\,\nabla^2u\quad+\frac{\mu}{\sigma}\int_0^R\int_0^{2\pi}(-ur)\,\mathrm{d}\phi\,\mathrm{d}r$$

[acceleration $\propto$ restoring force due to tension + restoring force due to excess pressure in bowl]

where c is the velocity of transverse waves on the skin, σ is the mass per unit area of the skin and μ is a parameter associated with the air in the bowl of the drum.

In the lowest mode of oscillation, the displacement is a function of radius r only (figure 9.15).

(a) Show that the equation for the lowest mode is

$$\frac{\partial^2u}{\partial t^2}=c^2\left(\frac{\partial^2u}{\partial r^2}+\frac{1}{r}\frac{\partial u}{\partial r}\right)-\frac{\mu}{\sigma}\int_0^R u2\pi r\,\mathrm{d}r.$$

[Appendix 5]

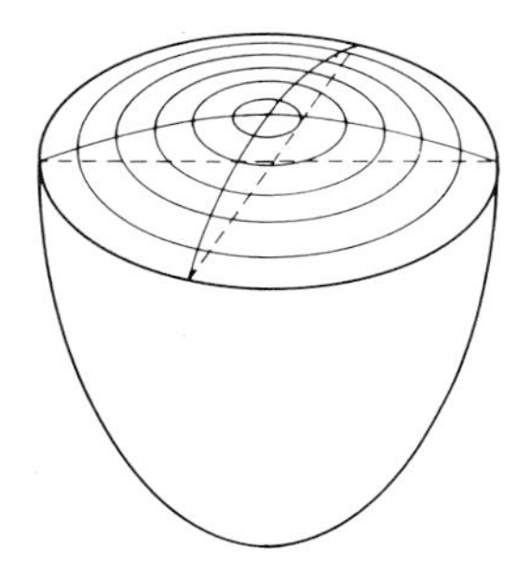

Figure 9.15 Lowest mode of vibration of a kettledrum.

(b) Use $u = U(r) \cos \omega t$ to separate this equation, and hence obtain the eigenvalue equation

$$\frac{\mathrm{d}^2 U}{\mathrm{d}r^2} + \frac{1}{r}\frac{\mathrm{d}U}{\mathrm{d}r} + k^2 U = \frac{2\pi\mu}{c^2\sigma}\int_0^R U r \, \mathrm{d}r$$

where

$$k = \omega / c.$$

(c) Use figure 9.15 (and 'U'' is finite' rule) to write down the boundary conditions for U, and hence show that the TF is

$$U^* = A\left(1 - \frac{r^2}{R^2}\right).$$

Write down the value of r for the collocation point.

(d) Show that

$$\int_0^R U^* r \, \mathrm{d}r = AR^2/4.$$

(e) Substitute this integral and U^* into the equation to obtain the residual equation, and collocate to obtain

$$k^2 = \frac{16}{3R^2}\left(1 + \frac{\pi\mu}{8c^2\sigma} R^4\right).$$

(f) Obtain the design formula

$$f = \frac{1}{2\pi}\left(\frac{16}{3}\right)^{1/2} \frac{c}{R}\left(1 + \frac{\pi\mu}{8c^2\sigma} R^4\right)^{1/2}.$$

(The second term in the bracket, involving μ, is the correction for the effect of air pressure. The correction term in real kettledrums is not small; the frequency shift can amount to more than 50%. It is not permissible in this case therefore to use the binomial approximation in the design formula for f.)

9.6

If the effect of stiffness is taken into account, the governing equation for the transverse vibrations of a wire is

$$\frac{\partial^2 u}{\partial t^2} = \frac{T}{\mu}\frac{\partial^2 u}{\partial x^2} - a\frac{\partial^4 u}{\partial x^4}.$$

[acceleration $\propto$ restoring force due to tension + restoring force due to stiffness]

The parameter a is a measure of the stiffness.

Putting $u = U\cos\omega t$ gives the eigenvalue equation

$$-a\frac{d^4U}{dx^4} + c^2\frac{d^2U}{dx^2} = -\omega^2 U$$

where $c = (T/\mu)^{1/2}$ is the velocity of transverse waves on the wire.

If the ends of the wire are wrapped around pins, as is usual in musical instruments, the boundary conditions at the two ends at $x = \pm l$ are $U = 0$ and $U'' = 0$. (Note that there are four boundary conditions because the equation is of fourth order.) The qualitative sketch of the lowest eigenfunction is shown in figure 9.16(a). (For completeness figure 9.16(*b*) shows the difference caused by clamping the ends between rigid jaws; the shape of the curve is modified near the ends, and a different TF would be needed to describe this curve.) The goal is to study the effect of stiffness on the frequency.

(a) First use the TF

$$U^* = A\left(1 - \frac{x^2}{l^2}\right).$$

Substitute this TF into the eigenvalue equation, and show that it is unsatisfactory because the residual equation does not involve the stiffness parameter a.

Figure 9.16 Lowest modes of vibration of a stiff string: (*a*) pinned; (*b*) clamped. The ends of the string are at $x = \pm l$.

(b) If $a = 0$ the governing equation reduces to the equation for a flexible string (equation (8.4)), whose exact lowest eigenfunction is

$$U^* = A\cos(\pi x/L) \qquad L = 2l.$$

Confirm that this satisfies all four boundary conditions for the stiff pinned string and that, since it matches the qualitative sketch in figure 9.16(*a*), it is a plausible TF. Substitute it into the eigenvalue equation, and show that it is in fact an exact eigenfunction if

$$\omega^2 = c^2 \left(\frac{\pi}{L}\right)^2 \left(1 - a\,\frac{\pi^2}{c^2 L^2}\right).$$

(When a/L^2 tends to zero, $\omega \to c(\pi/L)$, so long wires with low stiffness behave like completely flexible strings, as would be expected. On the other hand, if L is small the correction term in the brackets is large, so short wires behave more like stiff bars, again as expected physically.)

(c) Because the exact form of the eigenfunctions is sinusoidal, it is possible in this case to find the eigenvalues for the higher harmonics, using the eigenfunction for the nth harmonic

$$U^* = \cos\,(n\pi x/L).$$

Show that the ratio of the frequency of the nth harmonic to that of the fundamental is given by

$$\frac{f_n}{f_1} = n\left(1 + \frac{a(n^2-1)\pi^2}{2c^2 L^2}\right) \qquad a \text{ small.}$$

(For the nth harmonic to be strictly in tune with the fundamental, this ratio should be exactly equal to n. The result shows that the harmonics of a stiff wire tend to be increasingly 'sharp' (higher frequency than the ideal). This sharpening is characteristic of piano tone, which would lack its usual 'colour' if the stiffness were absent.)

Chapter 10

The QSTF method for conduction and diffusion equations

10.1 Introduction

Conduction/diffusion equations are PDEs containing the operator $\partial/\partial t$ but not $\partial^2/\partial t^2$. The common form is exemplified by the equation for heat balance (in unit time) in a volume element δV of unit heat capacity:

$$D\nabla^2\theta \quad + \quad G \quad = \quad \frac{\partial\theta}{\partial t}. \tag{10.1}$$

$$\left[\begin{matrix}\text{heat transported into } \delta V \\ \text{from adjacent elements}\end{matrix} + \begin{matrix}\text{heat generated} \\ \text{in } \delta V\end{matrix} = \begin{matrix}\text{heat stored} \\ \text{in } \delta V\end{matrix}\right]$$

The Laplacian operator in the transport term is often simplified to its one-dimensional form in the appropriate coordinates. The generation term G may be a function of the variables, but is more usually zero or a constant (see for instance equations (8.1) and (8.15)).

In §8.6 it was pointed out that the exact solution for θ in a typical practical problem consists of a combination of oscillatory eigenfunctions, but the solution curves are often simple and smooth (see figure 8.4 for example). Often these solution curves can be more usefully approximated by standard functions, and characterised by time constants or length constants as already explained for ODE. The QSTF method will be used in this chapter to obtain design formulae for these quantities in three practical problems of heat and mass transport.

10.2 Development of laminar flow in a pipe

10.2.1 Mathematical model

The physical problem considered here is the acceleration of liquid in a pipe in response to a change in pressure. The liquid is initially at rest, the pressure difference across the ends of the pipe being zero. At $t=0$, the pressure difference is suddenly increased to p. The velocity v parallel to the axis of the pipe increases as a consequence of the force of the pressure, starting from zero and eventually reaching a steady value; this suggests an approximately exponential rise of velocity. The goal is to check this hypothesis and to estimate a time constant for the process, if appropriate.

The governing equation for the force balance on an element of unit volume, assuming the flow is laminar rather than turbulent, is

$$\eta\left(\frac{\partial^2 v}{\partial r^2}+\frac{1}{r}\frac{\partial v}{\partial r}\right)+\quad\frac{p}{L}\quad=\quad\rho\frac{\partial v}{\partial t}. \tag{10.2}$$

[viscous force + pressure force = mass × acceleration]

Here η and ρ are the Newtonian viscosity and the density of the liquid, and p is the pressure difference between the two ends of the pipe (of length L and radius R).

The axial velocity $v(r, t)$ is a function of r, the distance from the axis, and of the elapsed time t. The equation is a conduction/diffusion equation similar to equation (10.1), with the Laplacian ∇^2 taking the appropriate form for cylindrical symmetry (appendix 5).

Since the equation is first order in time and second order in space, it will need three auxiliary conditions:

(*i*) *Initial condition*

Velocity is zero for all r at $t=0$

$$v(r, 0)=0. \tag{10.3}$$

(*ii*) *Boundary conditions*

Velocity at pipe wall ($r=R$) is zero for all time; often called the 'no-slip' condition

$$v(R, t)=0 \tag{10.4}$$

and at the fictitious boundary ($r=0$)

$$\left.\frac{\partial v}{\partial r}\right|_{(0,t)}=0 \tag{10.5}$$

to ensure that the singular term $(1/r)(\partial v/\partial r)$ remains finite as $r\rightarrow 0$ (§9.6).

The condition at the final steady state, when the rate of change of v with time has become zero, is found by putting $\partial v/\partial t=0$ in equation (10.2) which gives

$$\frac{\eta}{r}\frac{\partial}{\partial r}\left(r\frac{\partial v}{\partial r}\right)+\frac{p}{L}=0$$

whose solution (see problem 5.2) is

$$v=v_\infty\left(1-\frac{r^2}{R^2}\right) \tag{10.6}$$

where v_∞ is the final steady-state velocity along the tube axis, given by

$$v_\infty = \frac{pR^2}{4\eta L}. \tag{10.7}$$

In the steady state, therefore, the velocity distribution is a parabola with its extremum at the pipe axis ($r=0$).

10.2.2 QSTF solution

Unlike the qualitative sketch for an ODE, which is a single curve showing the dependence of v upon x, the qualitative sketch needed for a conduction/diffusion equation is a set of several separate curves. Figure 10.1 shows the qualitative sketch set for this example. Each sketch shows the velocity profile v–r for successive times t.

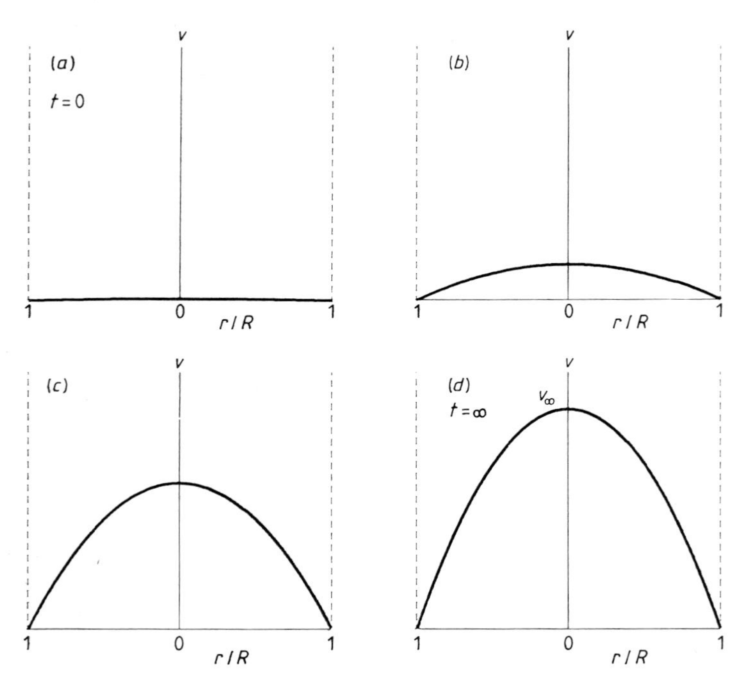

Figure 10.1 Qualitative sketches for the development of laminar flow profile.

At $t=0$ (figure 10.1(a)) the velocity is zero everywhere. After a short time (figure 10.1(b)) the velocity has increased near the centre, but must remain zero at the walls to satisfy the no-slip condition (10.4). The velocity profile continues to rise, always with a maximum at the centre (figure 10.1(c)), and eventually attains the final steady-state profile (figure 10.1(d)). These sketches, made using physical intuition together with the auxiliary conditions, suggest that the profile can always be approximated by a parabola, whose central height V increases with time from $V=0$ to $V=V_\infty$. The trial function for such parabolas is

$$V^*=V\left(1-\frac{r^2}{R^2}\right) \tag{10.8}$$

where V is the central velocity. Notice that unlike the ODE examples of Chapter 5, V is no longer a constant, but is an unknown function of time. Substituting this TF into the governing equation (10.2) gives

$$\frac{-4\eta V}{R^2}+\frac{p}{L}=\rho\left(1-\frac{r^2}{R^2}\right)\frac{\mathrm{d}V}{\mathrm{d}t}+\mathscr{R}. \tag{10.9}$$

(The differential is written as $\mathrm{d}V/\mathrm{d}t$ since V is a function of t only.)

This residual equation cannot be balanced for all r, because the parabolic profile assumed is only an approximation. Following the rule for parabolas and polar coordinates (§9.3), the equation is therefore collocated halfway across at $r/R=\frac{1}{2}$, to give

$$\frac{-4\eta V}{R^2}+\frac{p}{L}=\frac{3}{4}\rho\frac{\mathrm{d}V}{\mathrm{d}t}.$$

Since the equation no longer contains the variable r, it is an ODE describing the dependence of V upon t. Rearranged in the familiar form of a linear ODE, it is

$$\frac{\mathrm{d}V}{\mathrm{d}t}+\left(\frac{16}{3}\frac{\eta}{\rho R^2}\right)V=\frac{4}{3}\frac{p}{\rho L}.$$

The solution can be written down directly (§2.3) as an exponential rise from zero to V_∞, with a time constant

$$\tau=\frac{3}{16}\frac{\rho R^2}{\eta}=0.19\frac{R^2}{\nu} \tag{10.10}$$

where $\nu=\eta/\rho$ is the kinematic viscosity of the liquid. This is the design formula for the time constant of the growth of the height of the velocity profiles.

10.2.3 Discussion

The physical meaning of the TF solution is perfectly clear: if the velocity profiles are approximated by a series of parabola-like figures, their height rises exponentially to a final steady-state value given exactly by relation (10.6), with a time constant given by the design formula (10.10).

For comparison, the exact solution is

$$v=\frac{pR^2}{4\eta L}\left[\left(1-\frac{r^2}{R^2}\right)-8\sum_{n=1}^{\infty}\frac{J_0(\alpha_n r/R)}{\alpha_n^3 J_1(\alpha_n)}\exp\left(\frac{-\alpha_n^2\nu t}{R^2}\right)\right] \tag{10.11}$$

where α_n is the nth zero of the Bessel function $J_0(\alpha)$. This is an example of solutions as infinite series of eigenfunctions, which were discussed in §8.6. The higher eigenfunctions oscillate rapidly in space, and it is impossible to visualise this solution as a smooth curve without considerable computation.

The exact and TF solutions are compared in figure 10.2 for various values of t.

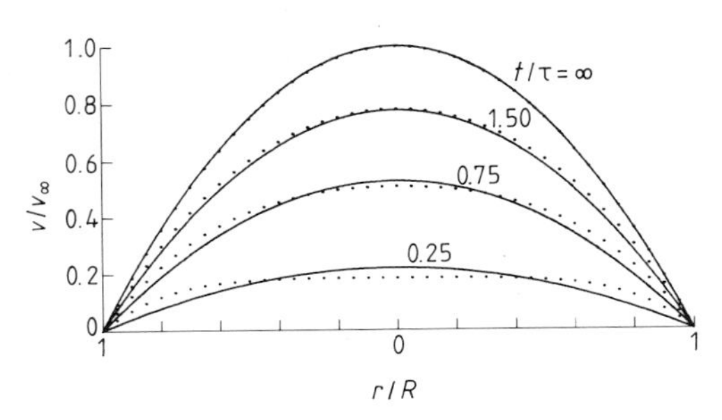

Figure 10.2 Comparison of TF (full curves) and exact (dotted curves) laminar flow profiles at various times.

The visual accord between the two sets of solution curves is excellent except for the earliest moments of the process, when the exact solution curve is flatter than the parabolic approximation. This is a consequence of the neglect of the higher decay modes in the solution, as discussed in §8.6.

10.3 Doping of a semiconductor

10.3.1 Mathematical model

One of the processes that takes place during the manufacture of semiconducting devices, such as transistors, involves the diffusion of an impurity into an initially pure wafer of silicon. The process, known as 'doping', is illustrated in figure 10.3. The silicon wafer is typically 50 mm in diameter and 0.2 mm thick. It rests on a solid tray.

The doping process begins when the impurity in the form of a gas is passed over the heated wafer. Impurity atoms start to diffuse into the wafer through the exposed top and sides. After a controlled doping time t, the impurity gas is

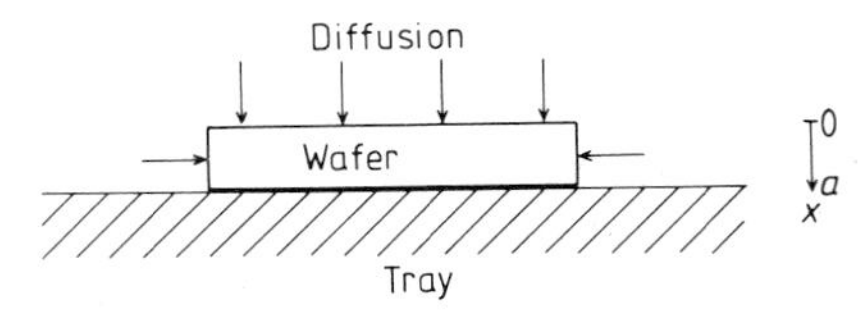

Figure 10.3 Diagram of a silicon wafer undergoing doping by diffusion of impurity gas atoms.

cut off to prevent further intake of impurities, and the wafer is cooled to prevent further diffusion inside.

The design problem is to determine the distribution of impurity atoms within the wafer as a function of diffusion time. The problem may be effectively regarded as one-dimensional by ignoring the very small amount of diffusion through the edges compared to that through the top face. The outer parts of the wafer are in any case discarded in practice.

The governing equation for the concentration c is the one-dimensional diffusion equation

$$\frac{\partial^2 c}{\partial x^2}=\frac{1}{D}\frac{\partial c}{\partial t}. \tag{10.12}$$

The auxiliary conditions are as follows:

(*i*) *Initial condition*
Initial concentration zero throughout wafer

$$c(x,0)=0. \tag{10.13}$$

(*ii*) *Boundary conditions*
Concentration maintained constant at top surface

$$c(0,t)=c_0 \tag{10.14}$$

and no diffusion into or out of the bottom surface

$$\left.\frac{\partial c}{\partial x}\right|_{(L,t)}=0. \tag{10.15}$$

A complete solution would yield an expression for the concentration at any depth and doping time; that is, $c(x,t)$ for $0\leqslant x\leqslant L$ and $0<t<\infty$. However, for most practical device purposes, the doping time is so short that impurities are only allowed to diffuse into a narrow layer near the top surface. The goal of the solution given here is therefore to find a design formula for the length constant characterising the concentration profile for such short doping times. (For longer doping times, see problem 10.2.)

10.3.2 QSTF solution

The qualitative sketches in figure 10.4 show the evolution of the concentration profile $c(x)$ from $t=0$ to $t=\infty$. In the initial state ($t=0$, figure 10.4(a)), the concentration is zero throughout the thickness of the wafer (condition (10.13)). The impurity gas has just been switched on, so that at the surface ($x=0$) the concentration is c_0, but there has been no time for the impurity atoms to penetrate beneath the surface. Some time later (figure 10.4(b) and (c)), the impurities have diffused a small distance from the surface, the concentration

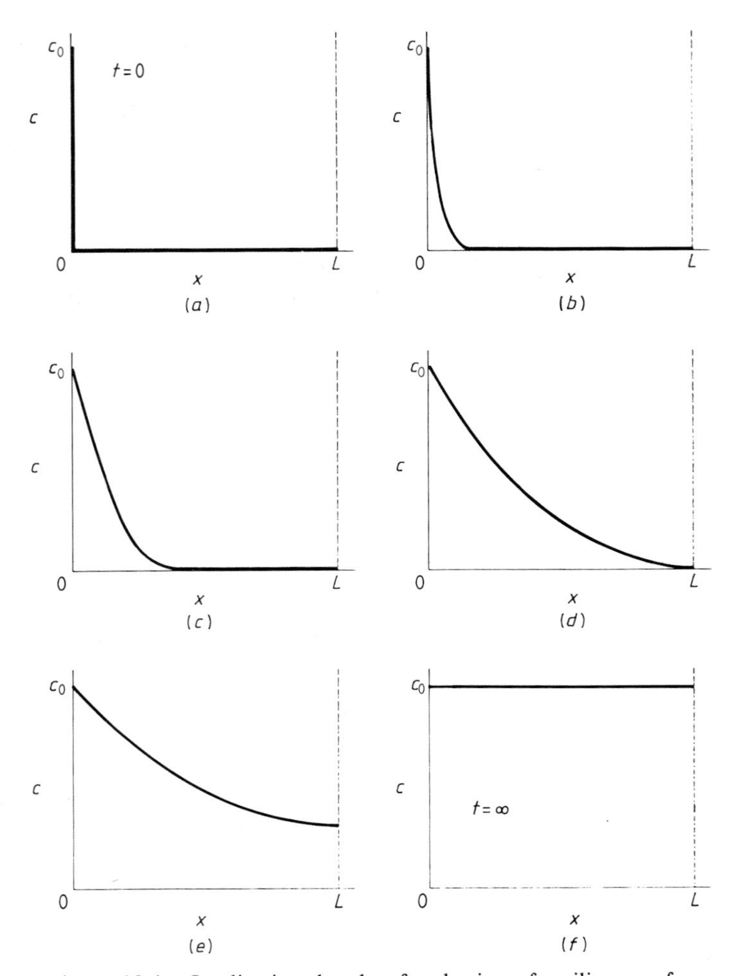

Figure 10.4 Qualitative sketches for doping of a silicon wafer.

falling smoothly to a long plateau. Eventually, as shown in figure 10.4(d), the impurity atoms just extend to the lower surface ($x=L$). Further diffusion raises the concentration at the lower surface above zero (figure 10.4(e)), until finally (figure 10.4(f)) the impurity atoms are distributed uniformly with concentration c_0 throughout the entire wafer. Note that all the sketches satisfy the boundary conditions (10.14) and (10.15), which state that the concentration is constant at $x=0$, and has zero gradient at $x=L$. A further point is that because the concentration is always increasing, $\partial c/\partial t$ is always positive everywhere in the wafer. It follows from equation (10.12) that $\partial^2 c/\partial x^2$ must also be positive everywhere, so the curves must bend away from the x-axis in the manner shown in the sketches, with the concavity facing upwards (see problem 2.9).

The plateaux in figure 10.4(*a*)–(*c*) show that the early stage of diffusion is a long-range problem (§4.4), for which an exponential TF is appropriate. The TF for each curve is therefore

$$c^* = c_0\, \mathrm{e}^{-x/\lambda} \tag{10.16}$$

in which λ is not a constant, but is a function of time. The goal is therefore to find a design formula for $\lambda(t)$, showing how the characteristic length in the TF (10.16) depends upon time.

Substitution of the TF (10.16) into equation (10.12) gives the residual equation

$$\frac{c_0}{\lambda^2}\,\mathrm{e}^{-x/\lambda} - \frac{c_0 x}{D\lambda^2}\,\mathrm{e}^{-x/\lambda}\,\frac{\mathrm{d}\lambda}{\mathrm{d}t} = \mathscr{R}. \tag{10.17}$$

The equation cannot be balanced for all x, so it is collocated, according to the rule for exponential TFS, at

$$\mathrm{e}^{-x/\lambda} = \tfrac{1}{2}$$

so that $x = \lambda \ln 2$ at the collocation point. With these substitutions and $\mathscr{R} = 0$, equation (10.17) becomes

$$1 - \frac{\lambda \ln 2}{D}\,\frac{\mathrm{d}\lambda}{\mathrm{d}t} = 0.$$

This is the ODE to determine $\lambda(t)$, which can now be solved exactly. By separation,

$$\lambda\, \mathrm{d}\lambda = \frac{D\, \mathrm{d}t}{\ln 2}.$$

Therefore

$$\tfrac{1}{2}\lambda^2 = \frac{Dt}{\ln 2} + \text{constant}.$$

The integration constant is determined by inspecting figure 10.4(*a*). This shows that when $t = 0$ the concentration falls to zero in an infinitesimal distance. The length constant λ is therefore zero when $t = 0$, and so the constant is also zero.

It follows that

$$\lambda^2 = \frac{2Dt}{\ln 2}$$

and therefore

$$\lambda = 1.7(Dt)^{1/2}. \tag{10.18}$$

This is the required design formula for the characteristic length as a function of the doping time.

10.3.3 Discussion

There is little to be gained, even numerically, by attempting a more exact solution, since the equation is a relatively crude model of the physical process; in particular, the diffusion 'constant' D may vary with concentration c. The main use of the model, for which the TF solution is perfectly adequate, is to provide understanding. The approximating exponential (10.16) shows the way in which the concentration profile extends into the wafer at any instant; longer doping times increase the length constant in accordance with formula (10.18) and the profiles push deeper and deeper into the wafer.

In addition, inverting formula (10.18) gives

$$t = 0.35\lambda^2/D$$

from which the doping time to achieve a required profile can be estimated; the square-law dependence then allows controlled adjustments to be made.

10.4 Freezing of peas

10.4.1 Mathematical model

One method of freezing vegetables very quickly to a low temperature is to immerse them in liquid nitrogen at its normal boiling point of 77 K. The model studied here is for the freezing of peas, which may be regarded as spheres having a radius R of a few millimetres. It will be assumed that the peas are initially at a uniform temperature following blanching, and that convective heat transfer to the liquid is sufficient to cool the outside of the peas to 77 K at the instant of immersion, $t = 0$. The final goal is to find a design formula for the time taken for the temperature at the centre of the peas to fall to a specified value close to 77 K. The first step is to find a design formula for the time constant of the fall in temperature at the centre.

The governing equation is the conduction equation in spherical coordinates (appendix 5)

$$\frac{\partial^2\theta}{\partial r^2} + \frac{2}{r}\frac{\partial\theta}{\partial r} = \frac{1}{D}\frac{\partial\theta}{\partial t} \tag{10.19}$$

where θ is the excess temperature above 77 K and D is the thermal diffusivity for pea material, assumed constant.

(i) Initial condition
Temperature initially at θ_0 throughout

$$\theta(r, 0) = \theta_0.$$

(ii) Boundary conditions
Outside temperature at 77 K

$$\theta(R, t) = 0$$

and second differential is finite (§9.6)

$$\left.\frac{\partial\theta}{\partial r}\right|_{(0,t)} = 0.$$

10.4.2 QSTF solution

The qualitative sketches are shown in figure 10.5. These show the temperature profiles evolving from the initial state (figure 10.5(*a*)) through a preliminary stage in which there is a plateau, and the temperature at the centre ($r=0$) remains very close to θ_0 (figure 10.5(*b*)). This stage ends after a time t_1 (figure 10.5(*c*)), when the cooling just extends to the centre at $r=0$. In the second stage (figure 10.5(*d*)–(*f*)) the temperature at $r=0$ progressively falls from θ_0 to θ_f, the final temperature at the centre when the cooling process is stopped.

The two stages described here correspond to the two stages in the collapse of the decay modes discussed in terms of eigenfunctions in §8.6. For physical understanding, it is more fruitful to regard the two stages as a long-range (exponential TF) problem succeeded by a short-range (parabolic TF) problem. A complete approximate treatment of the problem would therefore first find the time t_1 by using an exponential TF to match the sketches in figure 10.5(*a*)–(*c*), in a manner similar to the semiconductor doping problem considered in the previous section. However, the discussion in §8.6 shows that the first stage is completed in a very short time compared with that taken for the second stage to approach completion. It is sufficient, therefore, for design purposes to ignore t_1 and calculate t_f by considering only the second stage, illustrated by figure 10.5(*c*)–(*f*).

The TF that matches these curves is the standard parabola

$$\theta^* = A\left(1 - \frac{r^2}{R^2}\right)$$

where A is a function of t only, initially equal to θ_0 and falling steadily towards zero from $t=t_1$.

Substitution of the TF into equation (10.19) leads to the residual equation

$$\frac{-6A}{R^2} = \frac{1}{D}\left(1 - \frac{r^2}{R^2}\right)\frac{\mathrm{d}A}{\mathrm{d}t} + \mathscr{R}. \quad (10.20)$$

Collocation at $r/R = \frac{1}{2}$ gives the equation

$$\frac{\mathrm{d}A}{\mathrm{d}t} + \left(\frac{8D}{R^2}\right)A = 0.$$

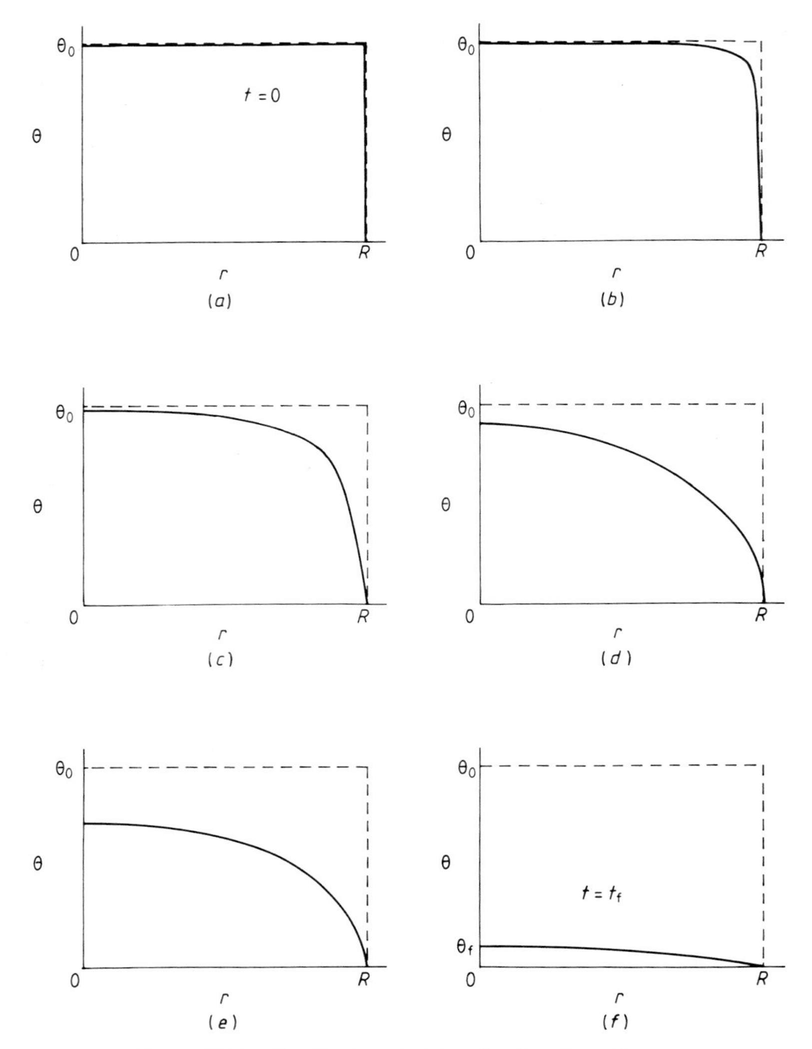

Figure 10.5 Qualitative sketches for freezing of peas.

This is an ODE 1 (§2.3) for A, the temperature at the centre. The solution is the exponential fall

$$A = \theta_0 \, \mathrm{e}^{-t/\tau} \tag{10.21}$$

where θ_0 is the initial excess temperature, and the design formula for the time constant is

$$\tau = \frac{R^2}{8D}. \tag{10.22}$$

10.4.3 Discussion

The final goal is to obtain an estimate of the freezing time t_f. This could be defined, for example, at the time for the temperature at the centre to fall by 90% of the complete change, which is from about 300 K to 77 K; this would reduce the temperature at the centre to about 100 K.

From table 3.2, the time to complete 90% of an exponential change is about 2.5τ. From the result (10.22), the final design formula for the freezing time is then

$$t_f = \frac{0.3R^2}{D}. \tag{10.23}$$

For $D = 10^{-6}\ \mathrm{m^2\ s^{-1}}$ and $R = 5$ mm, the typical freezing time is predicted to be about 8 s.

The reader should not be deceived by the simplicity of the result, nor the quick and simple way in which it was obtained. The formal solution is more involved and the result

$$\theta = \theta_0 \left[1 + \frac{2R}{\pi r} \sum_{n=1}^{\infty} \frac{(-1)^n}{n} \sin\left(\frac{n\pi r}{R}\right) \exp\left(-Dn^2\pi^2 t/R^2\right) \right]$$

is almost useless for physical understanding; its apparent exactness is of course illusory, since neither D nor R are precisely constant. Figure 10.6 shows graphs

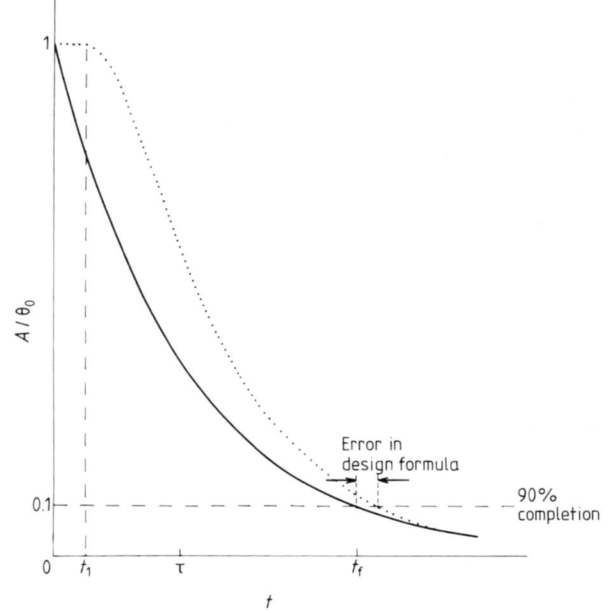

Figure 10.6 Comparison of TF solution (10.21) (full curve) and the exact solution (dotted curve) for the temperature at the centre of a freezing pea.

of the centre temperature as functions of time, according to the exact solution (dotted curve) and the TF result (10.21) (full curve).

The discrepancy for short times is caused by neglecting the first stage of cooling. As expected, this does not affect the good agreement between the approximate TF solution and 'exact' solutions near the completion of the process, which is close enough for design purposes.

Problems

10.1

In a timber yard, long wooden boards are seasoned by storing them so that they dry out by evaporation of water from their exposed faces, the ends and edges of the boards being sealed to prevent evaporation.

When the boards are first placed in position for seasoning ($t=0$), the water is distributed approximately uniformly throughout the thickness of the boards with a concentration c_0. By the argument of §8.6, the square initial profile quickly collapses to an approximately parabolic shape, having a maximum of approximately c_0 at the centre ($x=0$) and being zero at the faces ($x=\pm b$). During seasoning, evaporation maintains the concentration at zero at the faces of the board, while the concentration at the centre declines. A seasoning time t_s may be defined as the time for the concentration c at the centre to fall to one-tenth of its initial value c_0.

The governing equation for the concentration is

$$\frac{\partial^2 c}{\partial x^2}=\frac{1}{D}\frac{\partial c}{\partial t}$$

where D is the diffusion constant.

(a) Make qualitative sketches of the concentration profiles at various times from $t=0$ to $t=\infty$. [§10.2]

(b) Write down a TF for the declining parabolic profiles, containing an amplitude C which is an unknown function of time.

(c) Substitute, collocate to obtain an ODE for C and hence derive a design formula for the seasoning time.

(d) In the trade, there is an old rule that the seasoning time is one year for every inch of thickness, which works quite well for boards between 2 and 4 inch thick. Combine this information with your design formula to suggest a more accurate rule for boards of any thickness.

10.2

In certain applications of semiconductors, long diffusion times are used to obtain an almost uniform concentration of impurities throughout the thickness of the wafer (see figure 10.4(e) and (f)).

(a) Set up a parabolic TF for the later stage of the doping process discussed in §10.3, ensuring that it satisfies the boundary conditions $c(0)=c_0$ and $c'(L)=0$. [The problem is analogous to the freezing of a pea, discussed in §10.4]

(b) Substitute the TF into the governing equation (10.12) to obtain the residual equation, and collocate to obtain an ODE for the concentration C_L at the bottom face of the wafer.

(c) Write down the time constant for the approach of C_L towards c_0.

(d) Assuming that C_L starts from 0 at $t=0$, estimate the time for it to reach $0.99c_0$.

(e) Compare the time found in (d) with the time taken to complete the long-range exponential first stage of the profiles, taken to be when $2\lambda=L$. [Figure 10.4(*d*) and inverted formula (10.18)]

(The time found in (e) is sufficiently short for it to be neglected in comparison with the time found in (d). This is another example of the short time taken for the higher eigenmodes of the exact solution to decay, discussed in §8.6. For design purposes, it is sufficient to neglect this stage of the diffusion process when aiming for uniform doping.)

10.3

Temperature changes at the surface of the Earth penetrate a short distance into the soil or rock, but at the bottom of mines there is no perceptible effect caused by daily or annual temperature cycles. To gain an understanding of the way in which periodic surface temperature changes penetrate into the Earth, it is useful to consider an idealised model in which the surface temperature is assumed to be a harmonic function of the form

$$\theta_s=\theta_0 \cos \omega t$$

where

$$\omega=2\pi/T$$

and T might be either a day (8.6×10^4 s) or a year (3.2×10^7 s), according to whether daily or annual temperature changes are considered.

The governing equation for the temperature as a function of depth x below the surface and time t is

$$\frac{\partial^2\theta}{\partial x^2}=\frac{1}{D}\frac{\partial\theta}{\partial t}.$$

(Curvature of the Earth's surface is negligible for the small distances involved.)

(a) Describe in words the physical meaning of the TF

$$\theta^*=A\cos(\omega t-kx)$$

which has been proposed for the problem. Here A is assumed to be an unknown function of x.

(b) Substitute the TF into the governing equation to obtain the residual equation.

(c) Balance the two terms containing $\sin(\omega t - kx)$ and $\cos(\omega t - kx)$ to obtain a pair of differential equations for the function $A(x)$ and the constant k.

(d) Write down the exact solution for A from the first of these equations [§2.3] and the formula for the length constant associated with the decay of amplitude.

(e) Substitute this solution for A into the second equation, and hence derive the design formula for k.

(f) Use the results of (d) and (e) to estimate the depths below which daily and annual surface temperature variations are reduced to less than 10%. (Take $D = 1.2 \times 10^{-6}\ \mathrm{m^2\,s^{-1}}$, typical for surface rocks and ice.)

(g) What would be the time lag between annual temperature oscillations at the surface and at the corresponding depth calculated in (f)?

10.4

The age of the Earth has been estimated from observations of the temperature gradient at the surface. In the earliest model, it was assumed that the Earth began at a uniform temperature u_0 (°C). The surface quickly cooled to its present value of about 0°C. It was also assumed that cooling has only penetrated a small fraction of the distance towards the centre of the Earth, so that, as in the previous example, the curvature of the Earth's surface may be neglected. The model further assumed that there are no sources of heat generation within the Earth, so that the conduction equation contains no generation term. The appropriate form of the conduction equation (10.1) relating the temperature u to the depth x below the surface and time t was therefore assumed to be

$$\frac{\partial^2 u}{\partial x^2} = \frac{1}{D}\frac{\partial u}{\partial t}$$

with the boundary conditions $u(0, t) = 0$ and $u(\infty, t) = u_0$ (u_0 unknown). Here u_0 is the goal parameter, and the problem is to find a design formula relating u_0 to the surface temperature gradient $u'(0, t)$.

(a) Make a set of qualitative sketches consistent with this model.

(b) Write down a TF that matches these sketches, containing a length 'constant' that is an unknown function of time.

(c) Derive a design formula for the length constant as a function of time. [§10.3]

(d) Differentiate the TF, and use the result of (c) to obtain an expression for the temperature gradient at the surface.

(e) Taking $D = 6 \times 10^{-7}\ \mathrm{m^2\,s^{-1}}$, the experimental value of the surface temperature gradient as $3.6 \times 10^{-2}\ \mathrm{K\,m^{-1}}$, and the age of the Earth established on other grounds to be $\sim 4.5 \times 10^9$ years, show that the model requires an implausibly high initial temperature for the Earth.

(Step (d) of this problem involves differentiation of an approximate TF solution, the danger of which was mentioned in §1.5, and again in problem 4.2, as it exaggerates errors. The consequence of this is that the numerical accuracy of the gradient is likely to be worse than the accuracy of the original design formula for the length constant. Nevertheless, the result is sufficient to expose the unsatisfactory nature of the model, which is a consequence of the neglect of heat generated in the Earth's crust by radioactivity.)

10.5

Helium gas is generated in the rocks of the Earth's crust as a product of the radioactive decay of elements such as thorium and uranium. It diffuses to the surface and escapes into the atmosphere. In the simplest model of this process, the governing equation for the concentration c of helium at a depth x below the surface, at time t after the rocks were laid down, is

$$\frac{\partial^2 c}{\partial x^2}+\frac{G}{D}=\frac{1}{D}\frac{\partial c}{\partial t}$$

where D is the diffusion constant and G the rate of helium generation per unit volume. Both D and G are assumed to be constant throughout the crust.

The initial condition is that the concentration was zero everywhere when the rocks were first laid down. The two boundary conditions are that the surface concentration is zero at all times, and that the concentration gradient at great depths is zero. The goal is to find a design formula for the concentration gradient at the surface, from which the rate of helium diffusion into the atmosphere may be calculated.

(a) The second boundary condition implies that the concentration profiles at great depths have plateaux; this means that as $x \to \infty$ all the derivatives of c with respect to x are zero. Use this to obtain an auxiliary condition relating $c(\infty)$ to t.

(b) Make a set of qualitative sketches consistent with the information given and the auxiliary condition deduced in (a).

(c) Write down the TF, which should contain a length 'constant' that is an unknown function of time.

(d) Substitute and collocate to obtain the following ODE for the length constant

$$\lambda\frac{\mathrm{d}\lambda}{\mathrm{d}t}=\frac{D}{\ln 2}-\frac{\lambda^2}{t\ln 2}.$$

(e) At first sight this equation looks difficult to solve. However, all the examples of one-dimensional diffusion equations studied so far have contained length constants for which $\lambda \propto (Dt)^{1/2}$. It seems sensible therefore to use the TF $\lambda=(kDt)^{1/2}$. Do this, and hence derive the design formula

$$\lambda=0.86(Dt)^{1/2}.$$

(A proportionality $\lambda \propto t^{1/2}$ is also suggested by noting that the singular term $\lambda^2/(t \ln 2)$ remains finite at $t=0$ provided that $\lambda^2 \propto t$.)

(f) Differentiate the TF to obtain an approximate formula for the surface concentration gradient. Hence find an expression for the rate at which the model predicts that helium escapes into the Earth's atmosphere, in terms of the age T of the rocks, the radius R of the Earth, D and G.

(The same caution must be attached to the result of (f) as in part (d) of the previous problem, since it involves differentiating the approximate solution. Again, however, the approximate solution is useful for revealing the essential physics of the process and for making rough estimates of the rate of helium production.)

Chapter 11

Extending the QSTF method

11.1 Unfamiliar equations

The presentation of the QSTF method throughout this book has been facilitated by arranging the examples and problems into chapters, each dealing exclusively with either a particular type of TF or a particular type of equation. When an equation arises in some new practical problem, it may be obvious that it belongs to one of these classes, and it can then be solved with physical understanding by following the technique of the corresponding chapter. Sooner or later, however, an equation will arise that does not fit into any of these familiar classes, and without some further guidance the reader may feel helpless.

It is impossible to anticipate every type of equation that may arise from mathematical models, still less to devise schemes of solution that can always be carried through to a successful conclusion. However, except in specialised subjects, such as wave mechanics or networks, the equations are most commonly either ordinary differential equations of low order, or second-order partial differential equations whose solutions can be expressed as products, as discussed in chapters 8–10. The only other common physical equations are *algebraic equations*, which contain no differentials. These are usually much more straightforward to solve, and are treated briefly in appendix 4 and the second part of appendix 3. In this chapter, we shall summarise the QSTF plan of attack for unfamiliar ODEs, with the object of extracting as much physical understanding as possible.

The plan may be summarised in six steps:

(1) Inspect the equation and try to classify it into one of the familiar types already treated. If this is not immediately possible, classify it by its order, and note any nonlinearities (§2.2), singularities (§9.5) and the possibility of oscillatory solutions in time or space (§2.7). Then classify the auxiliary conditions as initial or boundary conditions. Finally, check for an eigenvalue equation (oscillatory spatial equation with boundary conditions, implying standing-wave solutions). For familiar equations, consult the appropriate chapter. For unfamiliar equations, read on.

(2) Define the goal; this may be modified later (see step (5)).

(3) Make a qualitative sketch. (The detailed means for doing this will be listed in §11.5.)

(4) By comparing the sketch with graphs of known functions, choose a TF containing the goal or closely related parameters. Use should be made of appendix 1, which collects together graphs of some of the most useful nonstandard functions besides those of the standard functions.

(5) Define (or redefine) the goal in terms of a design formula, in the light of the sketch and TF.

(6) Substitute the TF and use collocation or harmonic balance to obtain the design formula.

Steps (2) to (6) have already been used repeatedly in solving the examples of previous chapters. However these examples were all chosen so that in step (4) the suitable TF was one of the standard functions. To tackle the widest range of equations, it is necessary to add a new step at the beginning (step (1)), since the solver must now make the preliminary classification of the equation.

The plan will be illustrated in this chapter by taking three examples of unfamiliar equations; this will serve also to show that the use of nonstandard TFs. These examples will between them illustrate all the resources that are used to make qualitative sketches, whose key role will be discussed in §11.5.

11.2 The current gain of a junction transistor

The current gain β of a junction transistor may be calculated from the concentration of minority carriers in the base region. The formula is

$$\beta = \frac{n'(L)}{n'(0) - n'(L)} \tag{11.1}$$

where $n(x)$ is the minority carrier concentration at a distance x from the emitter/base junction at $x=0$. The base/collector junction is at $x=L$ (figure 11.1).

The governing equation for n is

$$\frac{\mathrm{d}^2 n}{\mathrm{d}x^2} - \frac{n}{b^2} = 0 \tag{11.2}$$

with the boundary conditions

$$n(0) = n_0 \qquad n(L) = 0. \tag{11.3}$$

Here b is the 'diffusion length', a semiconductor material parameter assumed to be constant and known.

Equation (11.2) is a linear ODE 2 with constant coefficients (§2.4), and is nonoscillatory since the coefficient of n is negative. The goal is a design formula for β.

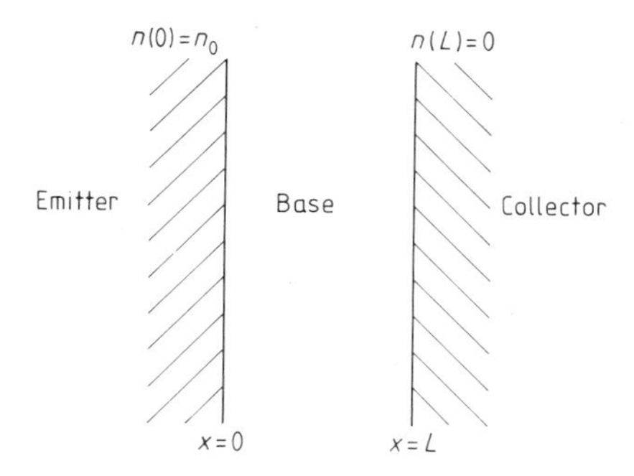

Figure 11.1 Schematic diagram of the base region of a junction transistor.

The conventional treatment of this problem consists of the following steps. First, the equation is solved exactly in terms of hyperbolic functions. Then the boundary conditions are imposed and the solution is differentiated to find $n'(0)$ and $n'(L)$. These values are then inserted into formula (11.1), and the expression expanded as a ratio of two power series. Finally the assumption is made that $L/b \ll 1$, and the power series simplified to give the design formula

$$\beta = 2\left(\frac{b}{L}\right)^2. \tag{11.4}$$

The steps are not in themselves difficult, but the cumulative working is lengthy and tedious.

The QSTF method is far simpler mathematically and a great deal more interesting physically. The starting point for the qualitative sketch is the pair of boundary conditions (11.3), which show that n falls from n_0 at $x=0$ to zero at $x=L$. Inspection of the equation (11.2) shows that, since

$$\frac{d^2n}{dx^2} = \frac{n}{b^2}$$

the second differential is always positive. The gradient therefore becomes more positive (or less negative) as x increases. The solution curve must accordingly fall smoothly, bending concave upwards (see problem 2.9) as indicated by the possible qualitative sketches in figure 11.2.

Practical transistors have large current gains ($\beta \gg 1$). In formula (11.1), the numerator is the concentration gradient at the end $x=L$, and the denominator is the change in gradient in going from $x=0$ to $x=L$. It follows that to obtain large values of β, this change in gradient must be small, so the qualitative sketch for a practical transistor must be very nearly a straight line (e.g. curve B in figure 11.2).

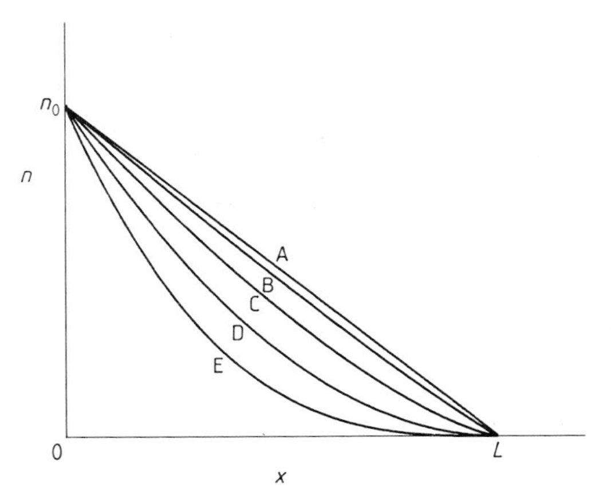

Figure 11.2 Qualitative sketches for the concentration of minority carriers across the base of a junction transistor. Curves A–E correspond to decreasing values of the parameter b in equation (11.2). Curve B corresponds to a practical transistor.

A suitable TF is therefore

$$n^* = n_0\left(1 - \frac{x}{L}\right) - ax\left(1 - \frac{x}{L}\right).$$

It may be confirmed that this satisfies the boundary conditions, and is made up of a straight line term plus a small correction, which is zero at each end and gives a downward bowing in the centre.

The disadvantage of this TF is that it does not contain the goal parameter β. Furthermore, it would have to be differentiated to find the gradients needed in formula (11.1). Fortunately, in this example, the TF itself is not needed; once it is established that the qualitative sketch is an almost straight line, β can be estimated directly from the equation. Integrating the equation between 0 and L gives

$$\int_0^L \frac{d^2n}{dx^2}\,dx = \frac{1}{b^2}\int_0^L n\,dx$$

or

$$n'(L) - n'(0) = \frac{1}{b^2}\int_0^L n\,dx.$$

The integral on the right is simply the area under the curve of n against x, and since the curve is almost straight this can be estimated directly as the area of a triangle of height n_0 and base L. Therefore

$$n'(L)-n'(0)=\frac{1}{b^2}\left(\frac{n_0L}{2}\right).$$

Again, since the line is almost straight, the gradient is to a good approximation equal to the average gradient, so that we may write

$$n'(L)=-n_0/L.$$

Inserting these expressions into the formula (11.1) for β gives the 'exact' design formula (11.4) directly.

11.3 The range of a high-energy particle

The *Bethe equation* describes the behaviour of an energetic particle, such as an electron, when it penetrates matter. The loss of energy of the particle by collisions with the atoms of the material is governed approximately (for $u \not< 1$) by the equation

$$\frac{\mathrm{d}u}{\mathrm{d}x}=-\frac{\ln(1+u)}{u}. \tag{11.5}$$

Here u is a dimensionless measure of the kinetic energy, and x is a dimensionless measure of the distance that the particle has penetrated into the matter. The value $x=0$ corresponds to the surface of the material, where the particle has the high initial energy $u(0)=u_0$. Numerically u_0 is typically 10^3 to 10^5.

Since the energy must fall smoothly from its initial value at the surface to near zero when the depth is very great, the first intuitive sketch might well be a falling exponential curve like figure 3.2. However, the equation is so different from anything previously considered in this book that it is wise to check this idea by using the equation to understand how the gradient changes as u falls.

The gradient at the surface, when $u=u_0$, is

$$u'(0)=-\frac{\ln(1+u_0)}{u_0}.$$

For the typical value $u_0=10^4$ this initial gradient has a value of about 10^{-3}. When u has fallen to half this initial value, the gradient is about -2×10^{-3}, still of the same order of magnitude. However, when u is close to zero, say $u=1$, the gradient is -0.69, which is three orders of magnitude steeper. The gradient is therefore always negative and its magnitude increases as u falls, that is as x increases. The final gradient is enormous compared with the initial gradient, and on a normal graph the curve will appear to be vertical when u approaches zero (figure 11.3).

This qualitative sketch has a very different shape of curve from the

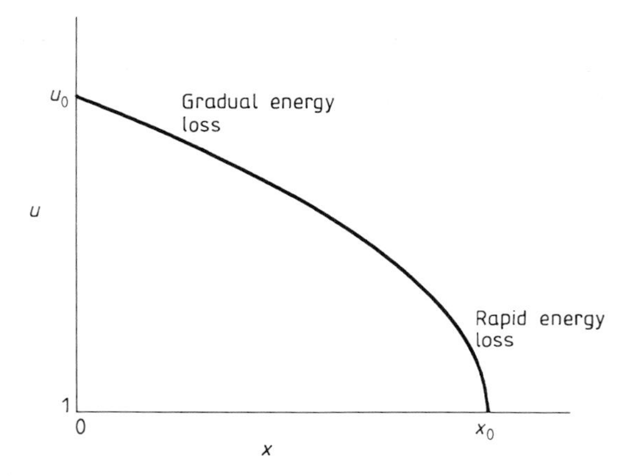

Figure 11.3 Qualitative sketch for the Bethe equation. The equation is only useful above $u=1$.

exponential, the curve bending increasingly downwards as x increases, rather than flattening out. Such a curve is characterised not by a length constant but by a *range*, defined as the distance at which the energy u becomes very small. Physically, the sketch suggests that the particle will retain a substantial fraction of its initial energy for a large part of the range, and will then lose this energy rather abruptly.

This conclusion, based on understanding via the qualitative sketch, is of considerable physical importance. The instruments used for detecting the particle depend upon the effects of its kinetic energy, and the final sharp fall in u means that instrumentally the particle will seem to disappear suddenly at a well defined range. By contrast, if the solution curve flattened out gently like an exponential, the range of detection would increase if a more sensitive instrument were used, and the range would be an ill defined experimental quantity.

The goal is to find a design formula for the range. The equation is a nonlinear ODE 1 and we have established that an exponential TF cannot be used; nor is a parabola suitable, since the curve does not have an extremum at $x=0$, but has a definite downward slope there.

A look through the curves in appendix 1 suggests that a *binomial function* of the form

$$u=u_0\left(1-\frac{x}{l}\right)^n$$

with n between 0 and 1, is a possible TF, since these functions have nonzero gradients at $x=0$ and a vertical gradient when $u=0$. These functions, however, contain two unknown parameters, n and l, and this means that two collocations will be required. The extra work that this would entail can be avoided by noting that ln u is a relatively slowly varying function of u. If it is regarded as approximately constant, the solution should not differ greatly from that of the simpler but analogous equation

$$\frac{\mathrm{d}u}{\mathrm{d}x}=-\frac{k}{u}$$

which is immediately soluble by separation (§2.2)

$$u\,\mathrm{d}u=-k\,\mathrm{d}x$$

$$\int u\,\mathrm{d}u=-\int k\,\mathrm{d}x+c$$

$$\tfrac{1}{2}u^2=-kx+c.$$

Using $u=u_0$ at $x=0$, we obtain the value of c:

$$\tfrac{1}{2}u_0^2=c$$

so

$$\tfrac{1}{2}u^2=\tfrac{1}{2}u_0^2-kx$$

whence

$$u=u_0\left(1-\frac{2kx}{u_0^2}\right)^{1/2}.$$

This is of the binomial form, suggesting that the argument by analogy is along the right lines, and that the constant n is approximately $\frac{1}{2}$. A binomial TF with $n=\frac{1}{2}$ should then be a good match to the exact solution curve. The TF is therefore

$$u^*=u_0\left(1-\frac{x}{x_0}\right)^{1/2}.$$

The only unknown quantity in the TF is x_0; since $u^*=0$ when $x=x_0$, the constant x_0 is the physical range of the particle, which is the goal of the problem.

Substitution of this TF into the governing equation (11.5) gives the residual equation

$$-\frac{u_0}{2x_0(1-x/x_0)^{1/2}}=-\frac{\ln[u_0(1-x/x_0)^{1/2}+1]}{u_0(1-x/x_0)^{1/2}}+\mathscr{R}.$$

The +1 in the logarithm bracket is negligibly small, and may be neglected.

We do not know where to collocate, but we can easily show that the exact choice is not critical. Collocation at $x/x_0=\frac{1}{2}$ (halfway across) gives

$$-\frac{u_0}{2x_0}=-\frac{\ln[(\frac{1}{2})^{1/2}u_0]}{u_0}$$

or

$$x_0 = \frac{u_0^2}{2(\ln u_0 - 0.35)}.$$

Alternatively, collocation halfway down, that is when

$$(1 - x/x_0)^{1/2} = \tfrac{1}{2}$$

gives

$$x_0 = \frac{u_0^2}{2(\ln u_0 - 0.7)}.$$

Since $\ln u_0$ is typically about 10, the difference between the two design formulae is only about 3%, and the compromise design formula

$$x_0 = \frac{u_0^2}{2(\ln u_0 - 0.5)}$$

gives excellent results for the range when compared to computed results.

This example shows that, in dealing with unfamiliar equations, it is important to check that the intuitive sketch is consistent with the gradients and curvatures indicated by the equation. It also shows once again how a correct qualitative sketch can give valuable physical understanding. Finally, it shows that the precise collocation point is not critical, so long as the TF is well chosen.

11.4 The Schrödinger equation for the harmonic oscillator

The Schrödinger equation for a particle moving in a straight line under the action of a central force $-\omega^2 x$ per unit mass is

$$\frac{d^2\psi}{dx^2} + \frac{8\pi^2}{h^2}(E - \tfrac{1}{2}\omega^2 x^2)\psi = 0 \qquad (11.6)$$

where ψ is the wavefunction, a quantity that is finite for all x, E is the total energy of the particle, and h is the Planck constant. The boundary conditions are

$$\psi(\infty) = 0 \qquad \psi(-\infty) = 0.$$

The exact solution can be obtained in terms of Hermite polynomials, and is a standard problem in quantum mechanics. Since our object is to understand the QSTF approach to unfamiliar equations, we shall imagine that this equation is encountered without any previous experience of Schrödinger equations and their solutions. The equation is an ODE 2 of unfamiliar form, so the sign test (§2.6) must first be used to determine whether it has oscillatory or non-oscillatory solutions.

The new feature is that q, the coefficient of ψ, changes sign as x increases. When x is very small, the equation becomes

$$\frac{d^2\psi}{dx^2}+\frac{8\pi^2}{h^2}E\psi=0$$

and the positive coefficient of ψ indicates that the equation describes oscillations of ψ in space. However, when x is very large, the equation becomes

$$\frac{d^2\psi}{dx^2}-\frac{8\pi^2}{h^2}\left(\frac{\omega^2x^2}{2}\right)\psi=0. \tag{11.7}$$

The coefficient of ψ is now negative, so the equation is no longer oscillatory. To investigate the nature of the solution of equation (11.7), consider the analogous linear equation

$$\frac{d^2\psi}{dx^2}-k^2\psi=0$$

which has the solution (§2.4)

$$\psi=A\,e^{-kx}.$$

The coefficient of ψ in equation (11.7) is analogous to k^2 in the linear equation, which suggests the correspondence

$$k^2=\frac{8\pi^2}{h^2}\left(\frac{\omega^2x^2}{2}\right)$$

or

$$k=\frac{2\pi}{h}\,\omega x.$$

Putting this in the solution (11.8) gives for $A=1$

$$\psi=\exp\left(-\frac{2\pi}{h}\,\omega x^2\right)$$

as the nonoscillatory solution applicable when x is very large.

Such arguments by analogy must not be taken literally, but they are suggestive of the functional form at large x. This idea is further strengthened by noting that

$$\psi=e^{-kx^2} \tag{11.9}$$

satisfies the boundary conditions that ψ be zero at both $x=-\infty$ and $x=+\infty$.

Having determined that the solution of equation (11.6) is oscillatory when x is small, but probably behaves like e^{-kx^2} when x is large in either the positive or negative direction, we now make qualitative sketches of possible solutions. If E is large, the equation remains oscillatory up to quite large magnitudes of x, because the coefficient of ψ in the equation is maintained positive by E until x^2 becomes sufficiently large to dominate. Many oscillations are possible before the change to nonoscillatory behaviour sets in (figure 11.4(*a*)). As E becomes smaller, fewer oscillations are possible before the nonoscillatory decay begins, and the curves become progressively simpler (figure 11.4(*b*)–(*d*)). It is now clear that the figures represent standing waves which fit between the boundaries, and are therefore eigenfunctions. The peculiarity of these standing waves is their exponential tails; these are a consequence of the boundaries being at infinity. This complication is one of the reasons why it is so much more difficult to solve and understand the Schrödinger equation than many other eigenvalue equations.

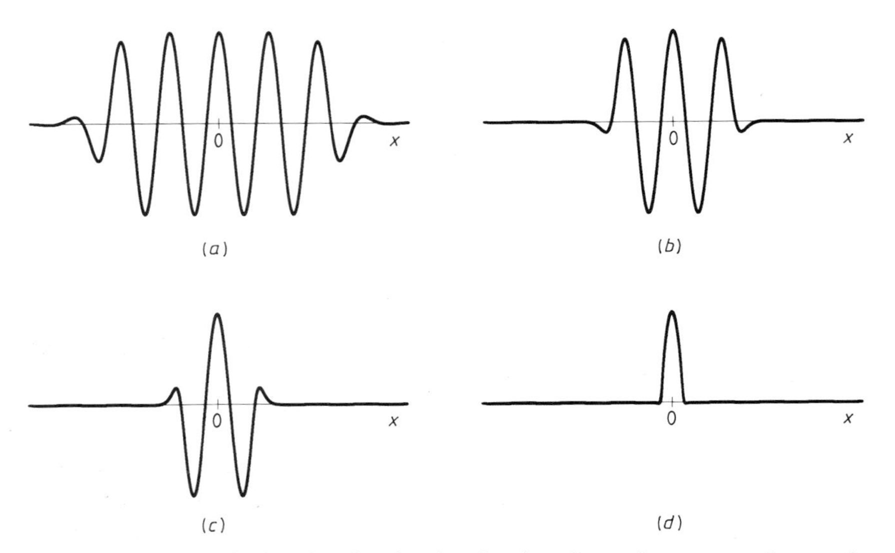

Figure 11.4 Qualitative sketches for the eigenfunctions of a quantum harmonic oscillator.

Figure 11.4(*a*), with many intermediate zeros, corresponds to a high-order eigenfunction. The lowest eigenfunction corresponds to the simplest curve (figure 11.4(*d*)). This cannot be matched by a standard function but looks like the *Gaussian curve* in appendix 1. The corresponding function is

$$\psi^* = \mathrm{e}^{-kx^2}$$

which has already been shown to have the correct form at $x = \pm\infty$, and is therefore a suitable TF for the lowest eigenfunction. Corresponding to this there will be a lowest eigenvalue of E, and our goal here is to find this eigenvalue E_0. Physically E_0 is of great interest since it is equal to the lowest possible energy of a quantum harmonic oscillator.

Substitution of this TF into (11.6) gives the residual equation

$$(-2k+4k^2x^2)\,\mathrm{e}^{-kx^2}+\frac{8\pi^2}{h^2}\left(E_0-\frac{\omega^2x^2}{2}\right)\mathrm{e}^{-kx^2}=\mathscr{R}.$$

The residual can be made zero for all values of x (indicating that the TF is an exact solution) if the coefficients of e^{-kx^2} and $x^2\,\mathrm{e}^{-kx^2}$ are made zero. This gives the two equations

$$-2k+\frac{8\pi^2}{h^2}E_0=0$$

and

$$4k^2+\frac{8\pi^2}{h^2}\left(-\frac{\omega^2}{2}\right)=0.$$

Elimination of k gives

$$\left(\frac{8\pi^2E_0}{h^2}\right)^2=\frac{4\pi^2\omega^2}{h^2}$$

whence

$$E_0=\frac{\omega h}{4\pi}.$$

Finally, by substituting $\omega=2\pi f$ we obtain the famous formula for the lowest possible energy of an oscillator of frequency f:

$$E_0=\tfrac{1}{2}hf.$$

The fact that harmonic oscillators could only have discrete values corresponding to the eigenvalues of E, and in particular that any oscillatory system cannot have zero energy, is one of the striking differences between the quantum theory and classical mechanics.

A further deduction from the treatment of the lowest energy of a quantum oscillator is the form of the higher eigenfunctions. The set of curves of the type shown in figure 11.4 have essential features in common with the *polyexponential functions* shown in appendix 1; the main difference is that the latter run from $x=0$ to $x=\infty$, whereas the curves in figure 11.4 run from $x=-\infty$ to $x=+\infty$. We can guess that for the harmonic oscillator the higher eigenfunctions are

$$\psi_n=P_n(x)\,\mathrm{e}^{-kx^2}$$

where $P_n(x)$ are polynomials of degree n. A detailed investigation confirms this hypothesis; $P_n(x)$ is called the nth *Hermite polynomial.*

It will be seen that, without using formal mathematics, but a succession of arguments by analogy, and the use of the qualitative sketch, the QSTF method gives a preliminary understanding of the quantum oscillator, together with some useful exact mathematical results.

11.5 The crucial role of the qualitative sketch

The emphasis on the systematic making of a qualitative sketch is a principal contribution of this book to the art of solving with physical understanding. It is defined in §1.3 as a sketch of the solution curve made without formally solving the equation; that is, by using physical rather than mathematical arguments. The method of constructing these sketches has been taught by its consistent use in almost every example. By analysing the way in which the sketches have been made in the examples, the technique can be seen to consist of finding clues to the likely shape of the curve from six distinct sources:

(a) The auxiliary conditions, initial, final and boundary, which fix the ends of the curve.

(b) Physical intuition, which suggests various possible or hypothetical curves between these endpoints, and then decides between them by checking against all the available data.

(c) The observation that most solution curves are either oscillations (or standing waves), long-range curves like exponentials, or short-range curves such as the parabola or binomial.

(d) Experimental evidence; for instance, in the transistor example of §11.2 the fact that $\beta \gg 1$ allows the 'almost straight line' curve to be chosen from other contenders.

(e) Analogy with simpler equations. This has constantly been used to discern oscillatory and standing-wave equations by means of the $q(u)$ test (§2.6) and in a more daring way in the quantum-mechanical harmonic oscillator example of §11.4.

(f) Inspection of the equation for direct information about gradients and curvatures (see problem 2.9 and, for instance, §§11.2 and 11.3).

This looking for clues, sifting the evidence and devising and testing hypotheses make up the essence of scientific work. It is what engineers and scientists do well, and can appreciate when it is well done by others. The importance of making the qualitative sketch is not just the result—a graph which could be plotted by other means—but the method of making it. Every time a scientist makes a qualitative sketch, he practises the art of scientific inference, rather than mathematical deduction. It is by the constant polishing of this art of inference and association of ideas that all new theories and inventions have been discovered; it is only later that they are dressed up in more formal clothes.

The other important aspect of the qualitative sketch is that it represents the furthest approach to solving an equation that can be made using purely verbal and graphical arguments. Many scientists and engineers, particularly those whose skills are predominantly practical, are accustomed to solving their problems using words, graphs and pictures, rather than by formal mathematics. Making a qualitative sketch by building it up by physical arguments is an excellent way of introducing the solution of a mathematical model to the widest possible audience, whether in a lecture or in a written paper. Both the nonspecialist and the nonmathematician, once they have seen the solution curve built up in this way, will be prepared to accept any formal or computed solutions that follow as reasonable and useful refinements of the outline solution already given.

11.6 Epilogue

The QSTF method, evolved to extract the greatest physical understanding, has been shown to yield useful solutions to a wide variety of differential equations, including many which are for practical purposes insoluble by more formal methods. The directness of the method makes it ideally suited to 'back-of-envelope' solutions of equations arising in technology, where economy of time and effort are usually more important than mathematical elegance or high accuracy.

Because of its emphasis on physical inference, rather than mathematical deduction, it offers a much broader view of mathematical models, often revealing features that are not easily discerned in formal or computed solutions. This broad view, which uses verbal and graphical as well as symbolic mathematical representations, is a stimulus to deeper thought about the model, and thence to creativity and invention. This widening of viewpoint makes the method the natural complement to computed solutions, the qualitative sketch and trial function providing the understanding, and the computation the accuracy and fine detail.

Finally, by concentrating first on physical arguments and confining mathematical operations to the simplest and most familiar possible, it opens up the understanding of many mathematical models to scientists whose skills are in experimental work rather than in mathematics. Mathematical models can therefore become a source of delight and scientific inspiration, since they can be solved, understood and explained using methods and language familiar to everyone engaged in practical scientific work.

References

Crandall S H 1956 *Engineering Analysis* (New York: McGraw-Hill)
Nayfeh A H and Mook D T 1979 *Nonlinear Oscillations* (New York: Wiley)

Appendix 1

Selected functions

Standard functions

(*i*) *Exponential decay*

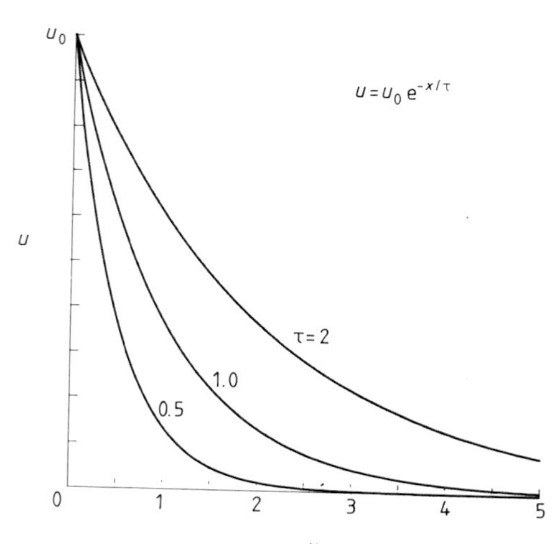

(*ii*) *Exponential rise*

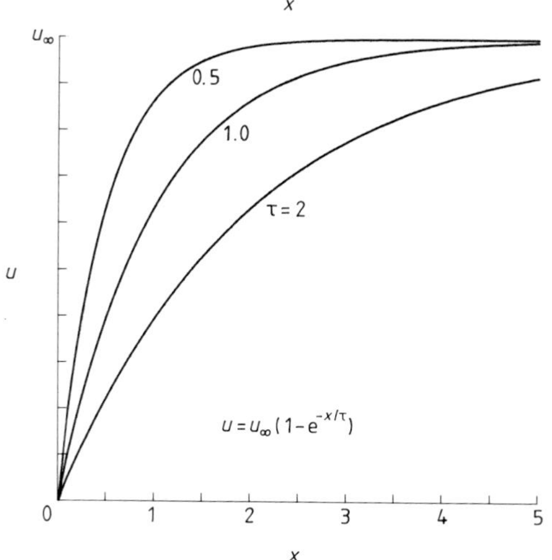

(iii) Exponential change

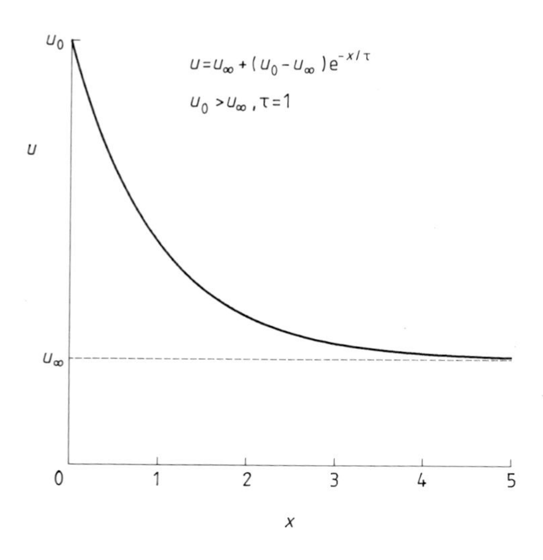

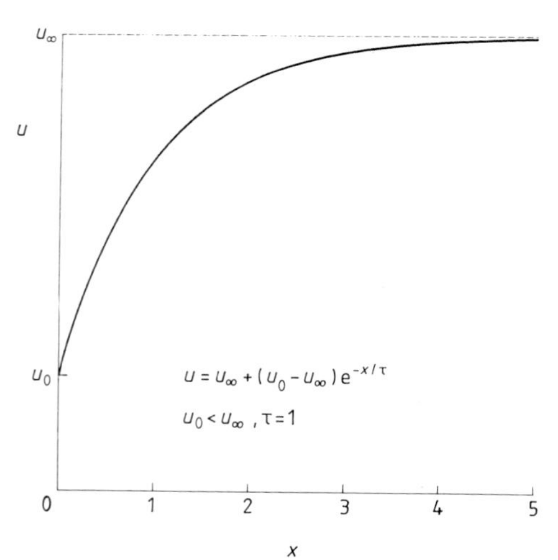

(iv) Parabola
(See also figure 5.1.)

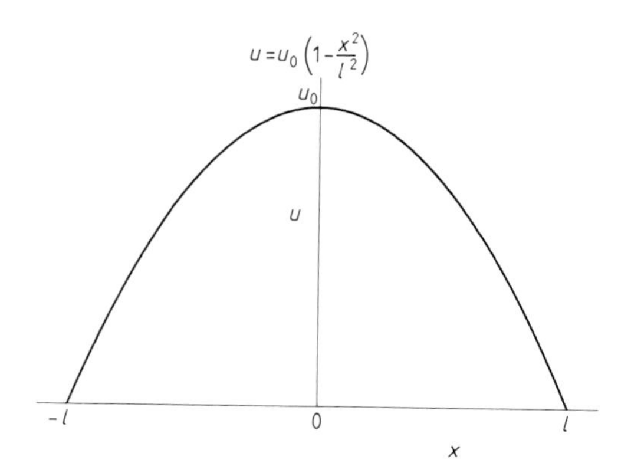

(v) Half-parabola

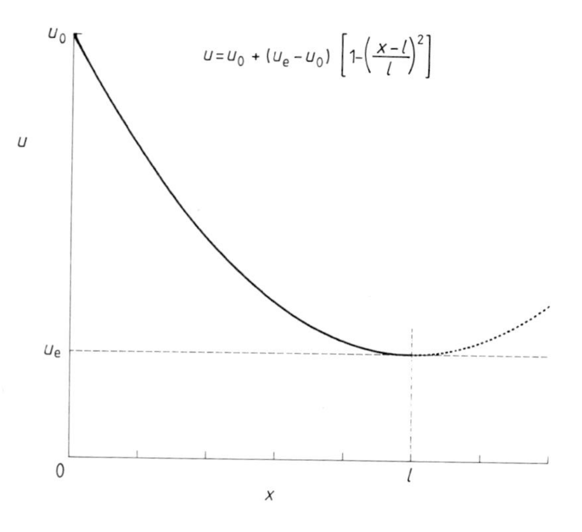

(vi) Cosine

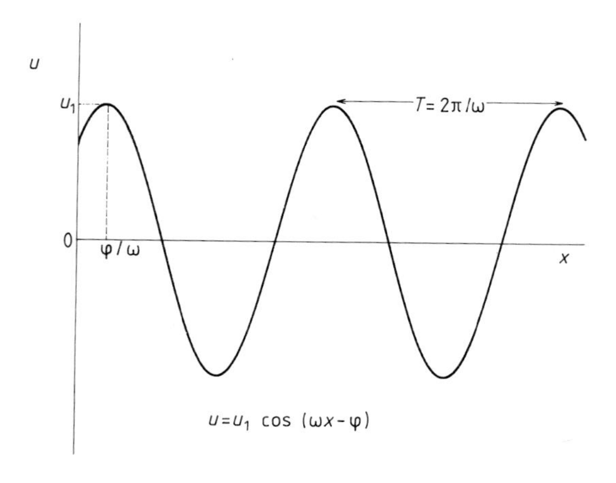

(vii) Shifted cosine

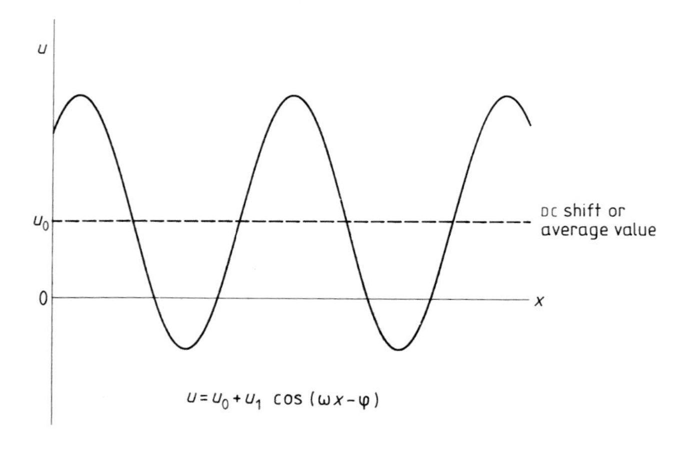

Other functions

(i) Quadratic

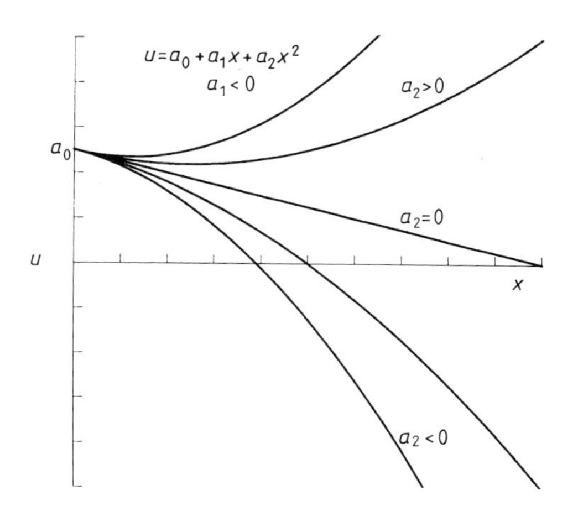

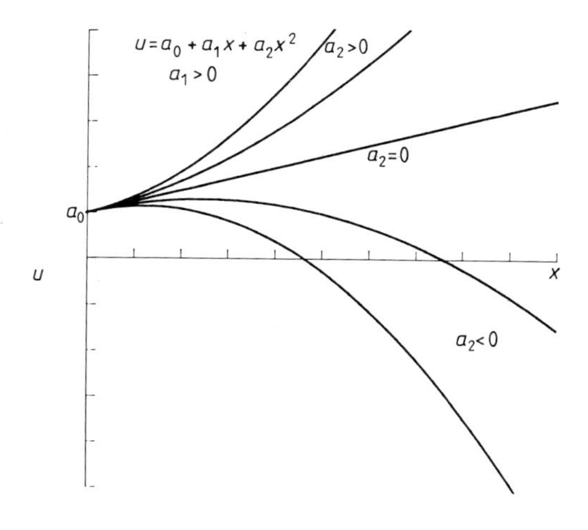

(ii) Gaussian

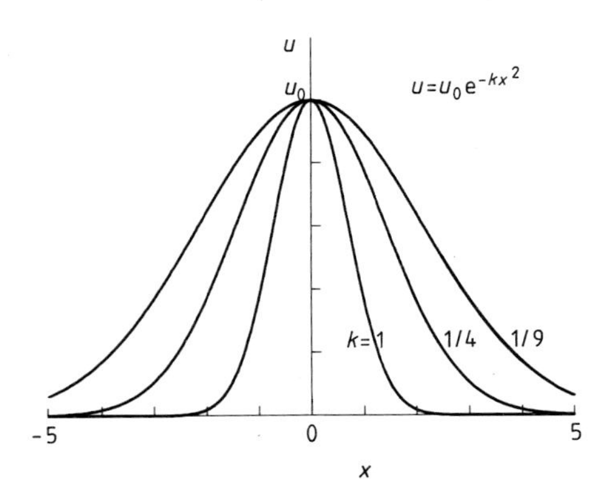

(iii) Damped cosine

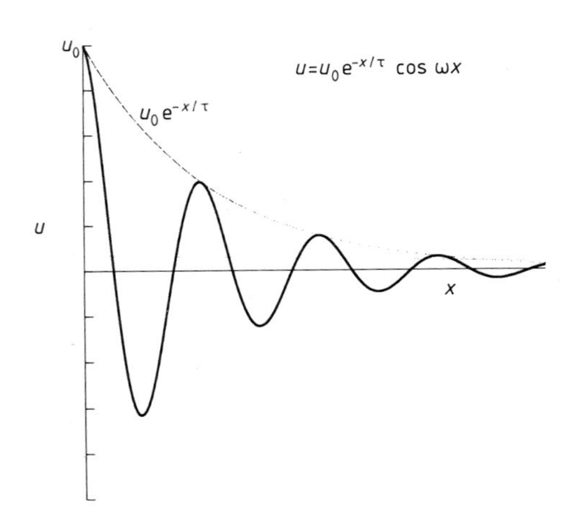

(iv) Binomial

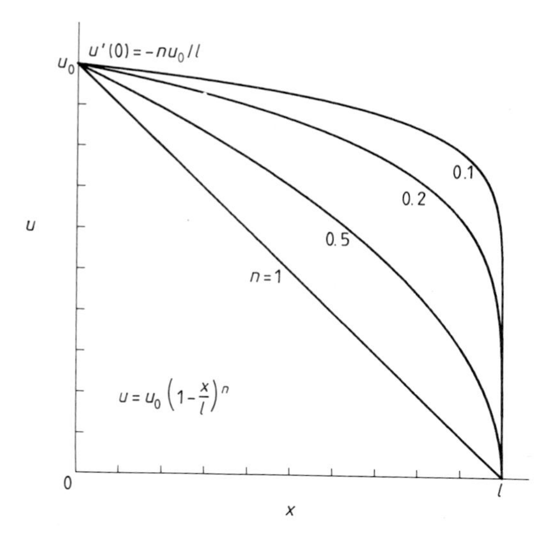
$u'(0) = -nu_0/l$
u_0
u
0.1
0.2
0.5
$n = 1$
$u = u_0\left(1 - \frac{x}{l}\right)^n$
0
l
x

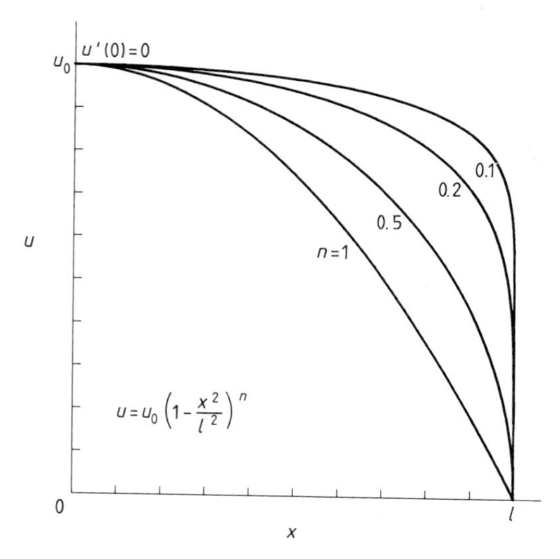
$u'(0) = 0$
u_0
u
0.1
0.2
0.5
$n = 1$
$u = u_0\left(1 - \frac{x^2}{l^2}\right)^n$
0
l
x

(*v*) *Polyexponentials*
General form

$$u = P_n(x)\mathrm{e}^{-kx}$$

where $P_n(x)$ is a polynomial of degree n. (N.B. The form of the curves is of n maxima and minima, followed by a final exponential decay to zero.)

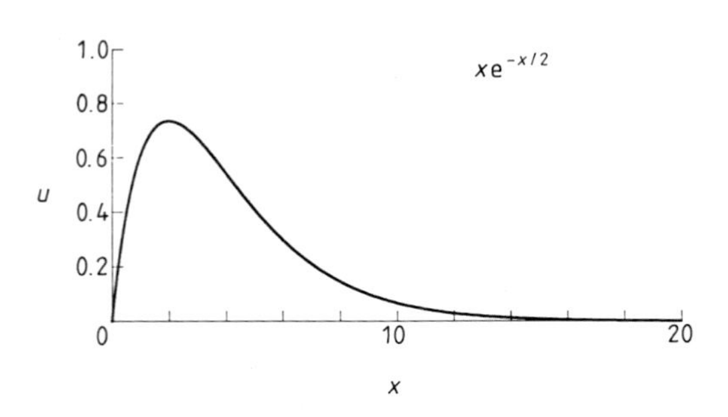

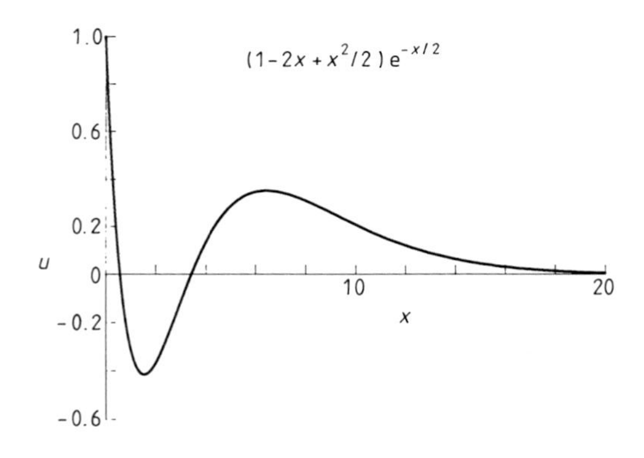

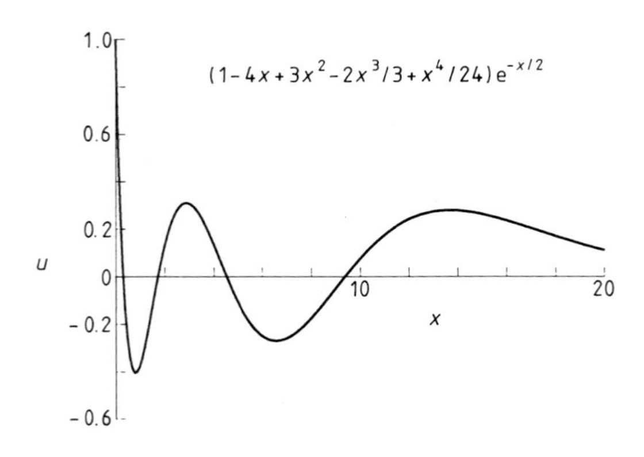

Appendix 2

Series

Maclaurin's expansion

$$u(x)=u(0)+u'(0)x+u''(0)\frac{x^2}{2}+\cdots+u^{(n)}(0)\frac{x^n}{n!}+\cdots$$

where $u'(0)$ denotes $\mathrm{d}u/\mathrm{d}x$ evaluated at $x=0$, $u''(0)$ denotes $\mathrm{d}^2u/\mathrm{d}x^2$ evaluated at $x=0$, and $u^{(n)}(0)$ denotes $\mathrm{d}^n u/\mathrm{d}x^n$ evaluated at $x=0$.

Binomial expansion

$$(1+x)^n=1+nx+\frac{n(n-1)}{2}x^2+\cdots+\frac{n!}{r!(n-r)!}x^r+\cdots.$$

Selected series

(Valid for all finite values of x unless otherwise shown.)

$$\frac{1}{1+x}=1-x+x^2-x^3+x^4+\cdots \qquad (x^2<1).$$

$$\frac{1}{1+x^2}=1-x^2+x^4-x^6+x^8+\cdots \qquad (x^2<1).$$

$$\mathrm{e}^{-x}=1-x+\frac{x^2}{2!}-\frac{x^3}{3!}+\frac{x^4}{4!}+\cdots$$

$$\mathrm{e}^{-x^2}=1-x^2+\frac{x^4}{2!}-\frac{x^6}{3!}+\frac{x^8}{4!}+\cdots.$$

$$\ln(1+x)=x-\frac{x^2}{2}+\frac{x^3}{3}-\frac{x^4}{4}+\cdots \qquad (-1<x\leqslant 1).$$

$$\sin x=x-\frac{x^3}{3!}+\frac{x^5}{5!}-\frac{x^7}{7!}+\cdots.$$

$$\cos x = 1 - \frac{x^2}{2!} + \frac{x^4}{4!} - \frac{x^6}{6!} + \cdots.$$

$$\tan x = x + \frac{x^3}{3} + \frac{2x^5}{15} + \frac{17x^7}{315} + \cdots \qquad (x^2 < \pi^2/4).$$

$$\sinh x = x + \frac{x^3}{3!} + \frac{x^5}{5!} + \frac{x^7}{7!} + \cdots.$$

$$\cosh x = 1 + \frac{x^2}{2!} + \frac{x^4}{4!} + \frac{x^6}{6!} + \cdots.$$

$$\tanh x = x - \frac{x^3}{3} + \frac{2x^5}{15} - \frac{17x^7}{315} + \cdots \qquad (x^2 < \pi^2/4).$$

$$J_0(x) = 1 - \frac{x^2}{2^2} + \frac{x^4}{2^2 \times 4^2} - \frac{x^6}{2^2 \times 4^2 \times 6^2} + \cdots.$$

$$\frac{\pi^{1/2}}{2} \operatorname{erf}(x) = x - \frac{x^3}{3 \times 1!} + \frac{x^5}{5 \times 2!} - \frac{x^7}{7 \times 3!} + \cdots.$$

Table A2.1 Coefficients of selected series.

	x^0	x^1	x^2	x^3	x^4	x^5	x^6
$(1+x)^n$	1	n	$\frac{1}{2}n(n-1)$	$\frac{1}{6}n(n-1)(n-2)$	–	–	–
$1/(1+x)$	1	-1	1	-1	1	-1	1
$1/(1+x^2)$	1		-1		1		-1
$\exp(x)$	1	1	1/2	1/6	1/24	1/120	1/720
$\exp(-x^2)$	1		-1		1/2		$-1/6$
$\ln(1+x)$		1	$-1/2$	1/3	$-1/4$	1/5	$-1/6$
$\sin x$		1		$-1/6$		1/120	
$\cos x$	1		$-1/2$		1/24		$-1/720$
$\tan x$		1		1/3		2/15	
$\sin^{-1} x$		1		1/6		3/40	
$\tan^{-1} x$		1		$-1/3$		1/5	
$\sinh x$		1		1/6		1/120	
$\cosh x$	1		1/2		1/24		1/720
$\tanh x$		1		$-1/3$		2/15	
$J_0(x)$	1		$-1/4$		1/64		$-1/2304$
$\frac{1}{2}\pi^{1/2} \operatorname{erf}(x)$		1		$-1/3$		1/10	

Appendix 3

Dedimensionalisation and reduction of parameters

There are some circumstances in which it is useful to reduce an equation to dimensionless form. We give here two examples illustrating a technique for dedimensionalisation that works well for both differential and algebraic equations; these examples will also show the advantages and disadvantages of the procedure.

Differential equations

The equation of a falling stone (§3.3) is

$$m\frac{\mathrm{d}v}{\mathrm{d}t}+Bv^2=mg. \tag{A3.1}$$

The corresponding dimensionless equation can be derived by following a strategy that is closely related to the TF method. We write

$$v=D_1u \qquad t=D_2x \tag{A3.2}$$

where D_1 and D_2 are as yet unknown constants that will later be adjusted to achieve our purpose. Substitution of these 'trial relations' into equation (A3.1) gives

$$\frac{mD_1}{D_2}\frac{\mathrm{d}u}{\mathrm{d}x}+BD_1^2u^2=mg.$$

Now remove the coefficient of the first term by multiplying the equation through by D_2/mD_1, so that

$$\frac{\mathrm{d}u}{\mathrm{d}x}+\left(\frac{BD_1D_2}{m}\right)u^2=\left(\frac{gD_2}{D_1}\right). \tag{A3.3}$$

To obtain the dimensionless form of equation (A3.3), choose D_1 and D_2 so that each of the brackets is equal to 1, by writing

$$\left(\frac{BD_1D_2}{m}\right)=1 \qquad \left(\frac{gD_2}{D_1}\right)=1.$$

Solving for D_1 and D_2 then gives

$$D_1 = \left(\frac{mg}{B}\right)^{1/2} \qquad D_2 = \left(\frac{m}{gB}\right)^{1/2} \tag{A3.4}$$

These are the required forms for the unknown constants D_1 and D_2 which make the brackets in equation (A3.3) equal to 1, and therefore reduce it to the dimensionless equation

$$\frac{\mathrm{d}u}{\mathrm{d}x} + u^2 = 1. \tag{A3.5}$$

This equation is mathematically much tidier than the original equation (A3.1). The most striking change is that it contains only the independent and dependent variables x and u, without any parameters. This reduction in the number of parameters greatly simplifies the presentation of a computed solution. Instead of a book full of tables or graphs to cover all the required permutations of the values of the three parameters, the solution can now be shown as a single table or a single graph of u against x. These can be expressed in terms of the original variables v and t, using the relations (A3.2) and (A3.4) to obtain

$$u = \frac{v}{D_1} = \left(\frac{B}{mg}\right)^{1/2} v$$

and

$$x = \frac{t}{D_2} = \left(\frac{gB}{m}\right)^{1/2} t.$$

It follows that by plotting $(B/mg)^{1/2}v$ against $(Bg/m)^{1/2}t$ rather than v against t, a single universal curve is obtained which shows the solution of the original modelling equation for every possible choice of parameters B, g and m, as well as of the independent variable t.

The maximum reduction in the number of parameters usually attainable by dedimensionalisation is three. Thus, when applied to the four-term equation

$$m\frac{\mathrm{d}v}{\mathrm{d}t} + Av + Bv^2 = T$$

the method leads to either

$$\frac{\mathrm{d}u}{\mathrm{d}t} + au + u^2 = 1$$

or

$$\frac{\mathrm{d}u}{\mathrm{d}t} + u + bu^2 = 1$$

or

$$\frac{\mathrm{d}u}{\mathrm{d}t}+u+u^2=c$$

according to which brackets are set equal to unity. However they are chosen, there is always a third bracket which cannot be made unity, and this appears as a parameter in the dimensionless equation. This reduction of four parameters to one is very useful in computer work, since graphical or tabulated solutions with one parameter can still be presented on one page.

Dedimensionalisation of differential equations is clearly of great value when the solutions are to be displayed graphically. It can also be useful at the classification stage of an equation (§11.1) by revealing more clearly the mathematical form of the equation. However, it is rarely of direct use in the working of the TF method. Since the goal is then a design formula showing the relation between the parameters, it is pointless to remove these parameters from the equation; it will be seen that it is one of the advantages of the TF method that it deals directly with the dimensional governing equation, dedimensionalisation being unnecessary and indeed undesirable.

Algebraic equations

These arise directly in mathematical models of systems in equilibrium; besides this, a relatively complicated algebraic equation occasionally arises as a result of the collocation of a differential equation in the TF method. Whatever their origin, the solutions of algebraic equations are best understood when presented graphically, so dedimensionalisation is a worthwhile first step for all but the simplest cases, in order to reduce the number of parameters.

As an example, consider the equation modelling electrically heated devices, such as soldering irons or electric fires or furnaces. A governing equation for the steady-state heat balance is

$$A(T^4-T_\mathrm{r}^4)+\quad B(T-T_\mathrm{r})\quad=\quad P.$$

$$\left[\text{radiation}\quad+\begin{array}{c}\text{convection}\\ \text{and conduction}\end{array}=\begin{array}{c}\text{electrical}\\ \text{power}\end{array}\right]$$

Here T is the absolute steady-state temperature for an electrical power P, T_r is the absolute room temperature and A and B are heat transfer coefficients for radiation and the combined effects of convection and conduction. The goal is to find a general graphical relation between T and P, which holds for any specified values of A, B and T_r; clearly this is impracticable without dedimensionalisation.

Before dedimensionalisation, the equation must be written in the conventional form, with the terms grouped in descending powers of T:

$$AT^4 + BT = (P + AT_r^4 + BT_r)$$

or

$$AT^4 + BT = P_e \tag{A3.6}$$

where P_e is the effective power; physically this is the electrical power plus the back-radiation, convection and conduction from the surrounding room. In any individual practical case, A, T_r and B are all specified, so P_e is known. Following the same dedimensionalisation procedure as before, write

$$T = D_1 u \qquad P_e = D_2 x. \tag{A3.7}$$

Substitute into equation (A3.6) to find

$$AD_1^4 u^4 + BD_1 u = D_2 x$$

or

$$u^4 + \left(\frac{B}{AD_1^3}\right) u = \left(\frac{D_2}{AD_1^4}\right) x. \tag{A3.8}$$

Set the bracketed terms equal to 1, so that

$$\left(\frac{B}{AD_1^3}\right) = 1 \qquad \left(\frac{D_2}{AD_1^4}\right) = 1$$

whence

$$D_1 = \left(\frac{B}{A}\right)^{1/3} \qquad D_2 = \left(\frac{B}{A}\right)^{1/3} \tag{A3.9}$$

and equation (A3.8) reduces to

$$u^4 + u = x. \tag{A3.10}$$

Combining relations (A3.7) and (A3.9) gives

$$u = \left(\frac{A}{B}\right)^{1/3} T \tag{A3.11}$$

$$x = \left(\frac{A}{B^4}\right)^{1/3} P_e = \left(\frac{A}{B^4}\right)^{1/3} (P + AT_r^4 + BT_r). \tag{A3.12}$$

Dimensionless algebraic equations such as (A3.10) can be solved graphically using a calculator. In appendix 4 the solution of equation (A3.10), and hence of the original modelling equation, is obtained as a single curve which covers every possible choice of input power P and parameters A, B and T_r.

Appendix 4

Solution of algebraic equations

Algebraic equations, by which we mean equations that do not contain differentials, commonly fall into one of two classes.

(a) Equations like

$$u = \tan u$$

or

$$u^3 + 2u - 1 = 0$$

which contain only a single variable (u in the examples). The goal of solving is to find the numerical value or values of u that satisfy the equation, as well as any physical restrictions, such as that u must be real and positive. The solution is a number or a set of numbers, so these equations have *numerical solutions.*

(b) Equations like

$$u = e^{-xu}$$

or

$$u^4 + u = x$$

which have two variables. For these the goal is not a number, but a *functional solution* showing the relationship between the dependent variable u and independent variable x. This relationship can be shown either by a mathematical formula or by a graph or table. Such equations often occur as the result of dedimensionalising governing equations that model steady-state systems (see appendix 3).

Numerical solutions

These are found very easily on a calculator, but to prevent errors it is wise to make a sketch first, so as to understand the number and approximate location of the possible roots. Consider, for example, finding the positive nonzero roots of

$$u^3 + 2u - 1 = 0.$$

A good way of locating the roots is to rearrange the equation so that the LHS and RHS are both functions that can be easily sketched. In the example

$$u^3 = 1 - 2u$$

and the functions on the LHS and RHS are sketched in figure A4.1. There is only one nonzero intersection of the curves, which shows that the root lies between 0 and 0.5.

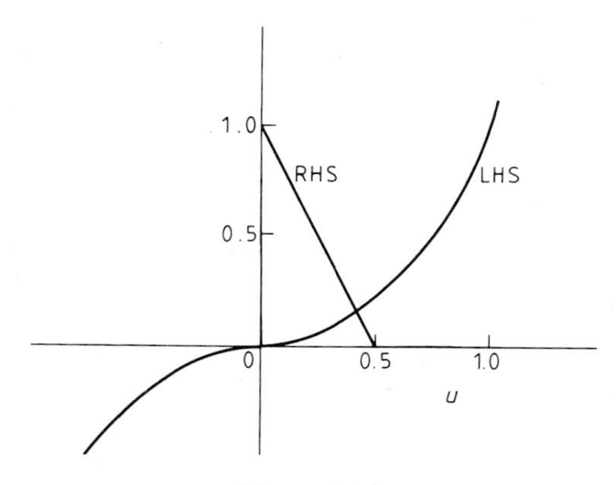

Figure A4.1

To use the calculator, the equation is next written in the form

$$f(u) = 0.$$

For the example

$$f(u) = u^3 + 2u - 1.$$

The calculator is programmed to evaluate this function for a succession of values of u entered manually. To start with, the function is evaluated for the outer limits of the root found from the sketch (steps 1 and 2 below).

Step	u	$f(u)$
1	0	-1
2	0.5	$+0.125$
3	0.2	-0.592
4	0.4	-0.136
5	0.45	-0.0089
6	0.46	$+0.017$
7	0.453	-0.0010

The first two entries show that for this example, if too low a value is guessed for u, $f(u)$ comes out negative; whereas too high a value for u makes $f(u)$ positive†. This information is used to make closer and closer guesses for the correct root of u. Step 3 shows that the next trial, $u=0.2$, makes $f(u)$ negative, so the guess is too low. Next $u=0.4$ is tried, still proving too low. Steps 5 and 6 show that u lies between 0.45 and 0.46, and the values of $f(u)$ suggest that u is nearer to 0.45 than 0.46. Step 7 shows that for $u=0.453$ the residual is only about 1/1000 of the magnitude of the typical term in the equation, so for most practical purposes this is the solution. The process can in principle be continued indefinitely to obtain any desired accuracy, but it is very rare to require more than 10 steps.

As a second example consider

$$\tan u = u.$$

Strictly speaking, this is a transcendental rather than an algebraic equation, but the goal is still a numerical solution, and the same method can still be applied.

Figure A4.2 shows sketches of the functions on the two sides of the equation. Like many equations containing periodic functions, there are infinitely many

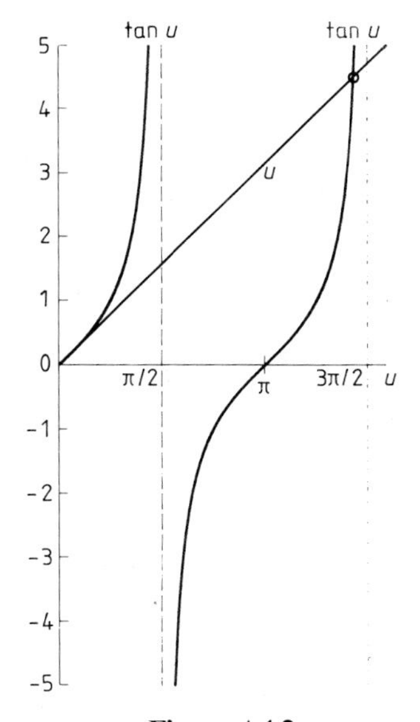

Figure A4.2

† Readers may recognise that $f(u)$ is the residual (§3.3.4), whose magnitude indicates the size of the error, and whose sign indicates which way to move to obtain a better trial number.

roots. Usually it is the first one or two nonzero positive roots that are required; the sketch shows the first of these to be situated at the intersection just below $u=3\pi/2\approx4.7$.

Because of the discontinuity in tan u at $u=3\pi/2$, it is advisable to start well below this figure and work cautiously upwards. For $f(u)=u-\tan u$ the first two entries below show that both guesses are too low. Steps 3 and 4 establish that the solution is between 4.4. and 4.5, and nearer the latter, which guides the subsequent guesses; step 9 suggests that $u=4.49341$ is the answer correct to six significant figures.

Step	u	$f(u)$
1	4.0	+2.8
2	4.2	+2.42
3	4.4	+1.3
4	4.5	−0.137
5	4.48	+0.25
6	4.493	+0.008
7	4.495	−0.032
8	4.4934	+0.0002
9	4.49341	−0.00001

Functional solutions

Graphical solutions showing the functional relation between variables are probably the most useful both for understanding and for design purposes. They can usually be obtained using a calculator and the device of *inverse tabulation.*

Consider the equation

$$u^4+u=x.$$

Regarded as an equation for u, this is quartic, and its solution either by computer or formally is unnecessarily tedious. Inverse tabulation simply means turning the equation round so that x is the subject:

$$x=u^4+u$$

and x can then be evaluated for successive values of u.

As usual, it is a good idea to understand the nature of the roots by sketching the functions on both sides of the equation. Figure A4.3 shows the graph of u^4+u against u, with a particular value of x drawn as a straight line across it. Whatever positive value is chosen for x there is only one intersection for which u is positive, and this is the solution we shall investigate.

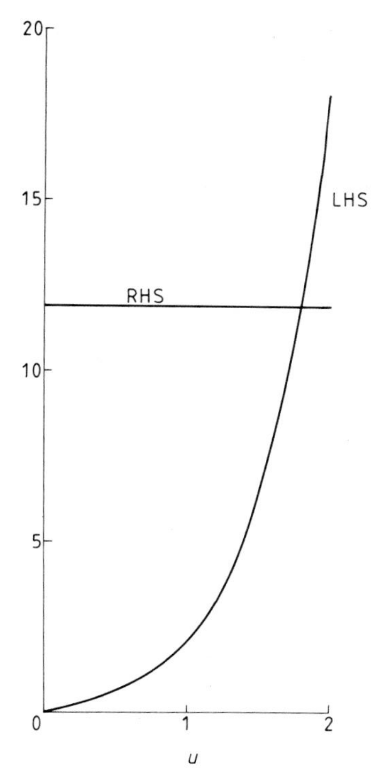

Figure A4.3

We tabulate x as a function of u as follows.

Entry	u	$x=u+u^4$
1	0	0
2	1	2
3	2	18
4	0.5	0.56
5	0.7	0.94
6	1.2	3.27
7	1.4	5.24
8	1.6	8.15
9	1.8	12.30

The first three entries establish that for $u=0$ to 2, x ranges from 0 to near 20, so these are suitable scales for the graph. The other six entries then fill in enough intermediate points to make a sufficiently accurate plot.

To present u as the variable dependent upon x, the graph must be plotted

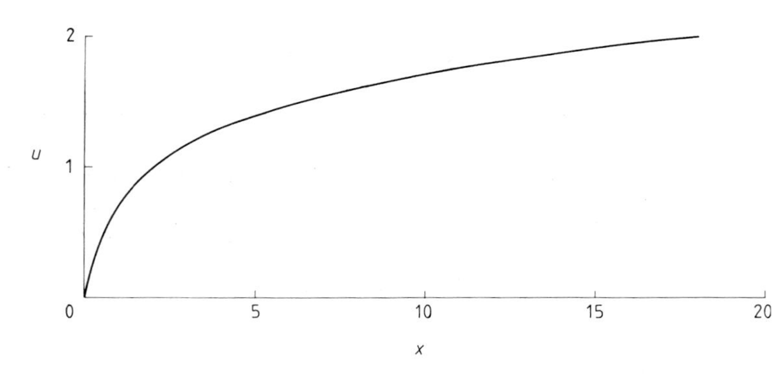

Figure A4.4 Universal curve for steady-state temperature of heated devices. See appendix 3 for definition of parameters.

with x along the horizontal axis and u along the vertical axis. The result is figure A4.4; this appears as if u has been calculated for various values of x, instead of the inverse process actually performed.

The equation

$$u^4 + u = x$$

is the dimensionless form of the governing equation for the equilibrium temperature of electrically heated apparatus, discussed in appendix 3. Figure A4.4 therefore answers the problem posed there, by providing a single curve relating the steady temperature to the input power for any values of the system parameters.

The method of inverse tabulation works well for a surprising number of otherwise tricky equations. For instance, it would be quite difficult to find the solution $u = f(x)$ of the equation

$$u = e^{-u/x}$$

directly; but it can be inverted to give

$$x = -\frac{\ln u}{u}$$

and inverse tabulation made of x for various values of u. This table can then be plotted, as in the previous example, to show how u (on the vertical axis) varies with x, thus giving a graphical presentation of the solution $u = f(x)$.

Appendix 5

Coordinate systems, Laplacian operator and important partial differential equations

Cartesian coordinates

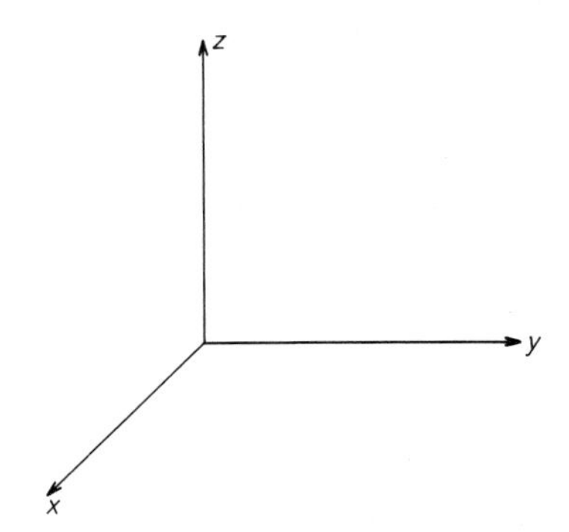

Laplacian

$$\nabla^2 u \equiv \frac{\partial^2 u}{\partial x^2} + \frac{\partial^2 u}{\partial y^2} + \frac{\partial^2 u}{\partial z^2}.$$

Cylindrical coordinates

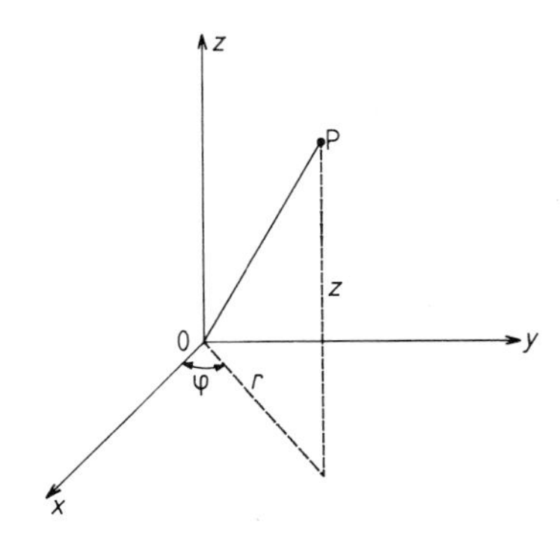

Laplacian

Full form

$$\nabla^2 u \equiv \frac{1}{r}\frac{\partial}{\partial r}\left(r\frac{\partial u}{\partial r}\right)+\frac{1}{r^2}\frac{\partial^2 u}{\partial \phi^2}+\frac{\partial^2 u}{\partial z^2}.$$

Cylindrical symmetry with no z dependence

$$\nabla^2 u \equiv \frac{\partial^2 u}{\partial r^2}+\frac{1}{r}\frac{\partial u}{\partial r}.$$

Spherical polar coordinates

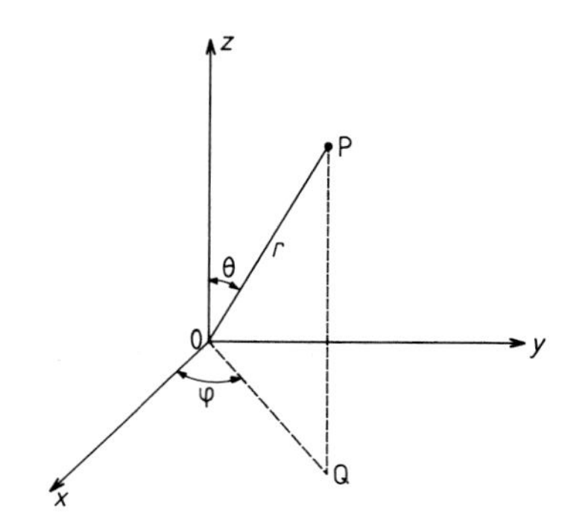

Laplacian

Full form

$$\nabla^2 u \equiv \frac{1}{r^2}\frac{\partial}{\partial r}\left(r^2\frac{\partial u}{\partial r}\right)+\frac{1}{r^2 \sin^2\theta}\frac{\partial^2 u}{\partial \phi^2}+\frac{1}{r^2 \sin\theta}\frac{\partial}{\partial \theta}\left(\sin\theta\frac{\partial u}{\partial \theta}\right).$$

Spherical symmetry

$$\nabla^2 u \equiv \frac{\partial^2 u}{\partial r^2}+\frac{2}{r}\frac{\partial u}{\partial r}.$$

Selected partial differential equations

Laplace

$$\nabla^2 u = 0.$$

Poisson

$$\nabla^2 u = f(x, y, z).$$

Helmholtz

$$\nabla^2 u + \lambda u = 0.$$

Conduction/diffusion

$$\nabla^2 u + g(x, y, z, t) = \frac{1}{D}\frac{\partial u}{\partial t}.$$

Wave

$$\nabla^2 u = \frac{1}{c^2}\frac{\partial^2 u}{\partial t^2}.$$

Schrödinger

$$\nabla^2 u + \frac{2m}{\hbar^2}[E - V(x, y, z)]u = 0.$$

Navier–Stokes

$$\nu\nabla^2 \boldsymbol{v} - \boldsymbol{v}\cdot \text{grad } \boldsymbol{v} + \boldsymbol{F}(x, y, z, t) = \frac{\text{grad } p}{\rho}\frac{\partial \boldsymbol{v}}{\partial t}.$$

Appendix 6

Trigonometric relations and Fourier series

Trigonometric formulae

$$\sin(A \pm B) = \sin A \cos B \pm \cos A \sin B.$$

$$\cos(A \pm B) = \cos A \cos B \mp \sin A \sin B.$$

$$\sin A + \sin B = 2 \sin\left(\frac{A+B}{2}\right) \cos\left(\frac{A-B}{2}\right).$$

$$\cos A + \cos B = 2 \cos\left(\frac{A+B}{2}\right) \cos\left(\frac{A-B}{2}\right).$$

$$\cos A - \cos B = -2 \sin\left(\frac{A+B}{2}\right) \sin\left(\frac{A-B}{2}\right).$$

$$A \cos(\omega t + \phi) = B \cos \omega t - C \sin \omega t$$

where $A^2 = B^2 + C^2$, $\phi = \tan^{-1}(C/B)$, and $B = A \cos \phi$, $C = A \sin \phi$.

$$\sin A \cos B = \tfrac{1}{2} \sin(A+B) + \tfrac{1}{2} \sin(A-B).$$

$$\cos A \cos B = \tfrac{1}{2} \cos(A+B) + \tfrac{1}{2} \cos(A-B).$$

$$\sin A \sin B = \tfrac{1}{2} \cos(A-B) - \tfrac{1}{2} \cos(A+B).$$

$$\cos^2 A = \tfrac{1}{2} + \tfrac{1}{2} \cos 2A.$$

$$\sin^2 A = \tfrac{1}{2} - \tfrac{1}{2} \cos 2A.$$

$$\cos A \sin A = \tfrac{1}{2} \sin 2A.$$

$$\cos^3 A = \tfrac{3}{4} \cos A + \tfrac{1}{4} \cos 3A.$$

$$\sin^3 A = \tfrac{3}{4} \sin A - \tfrac{1}{4} \sin 3A.$$

$$\cos^2 A \sin A = \tfrac{1}{4} \sin A + \tfrac{1}{4} \sin 3A.$$

$$\cos A \sin^2 A = \tfrac{1}{4} \cos A - \tfrac{1}{4} \cos 3A.$$

Fourier series

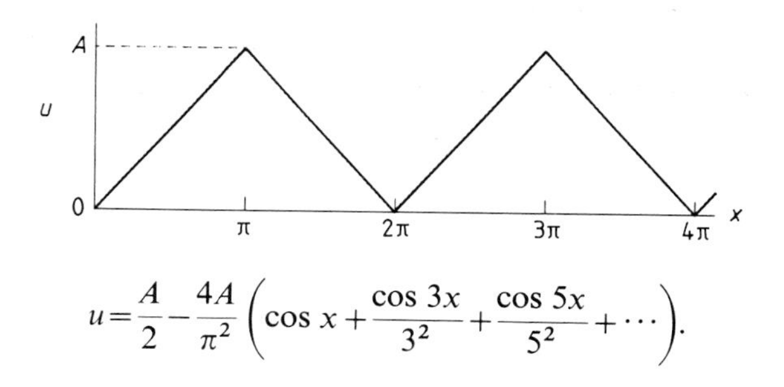

$$u=\frac{A}{2}-\frac{4A}{\pi^2}\left(\cos x+\frac{\cos 3x}{3^2}+\frac{\cos 5x}{5^2}+\cdots\right).$$

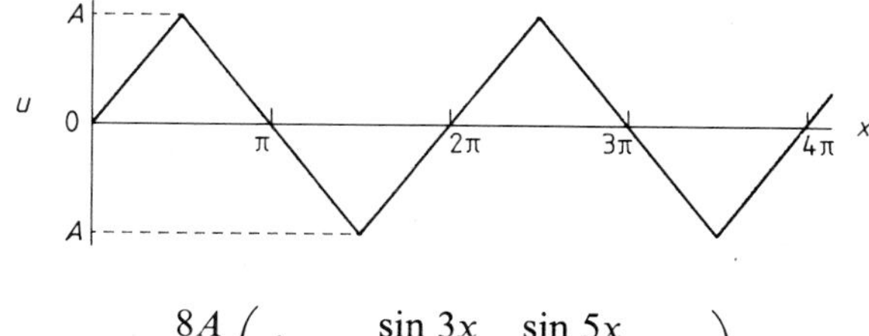

$$u=\frac{8A}{\pi^2}\left(\sin x-\frac{\sin 3x}{3^2}+\frac{\sin 5x}{5^2}+\cdots\right).$$

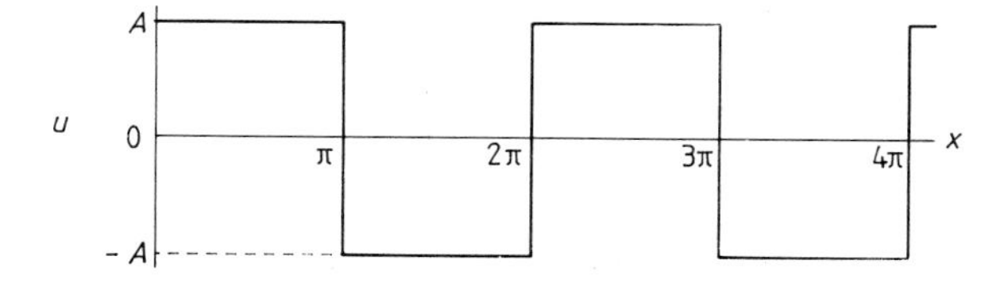

$$u=\frac{4A}{\pi}\left(\sin x+\frac{\sin 3x}{3}+\frac{\sin 5x}{5}+\cdots\right).$$

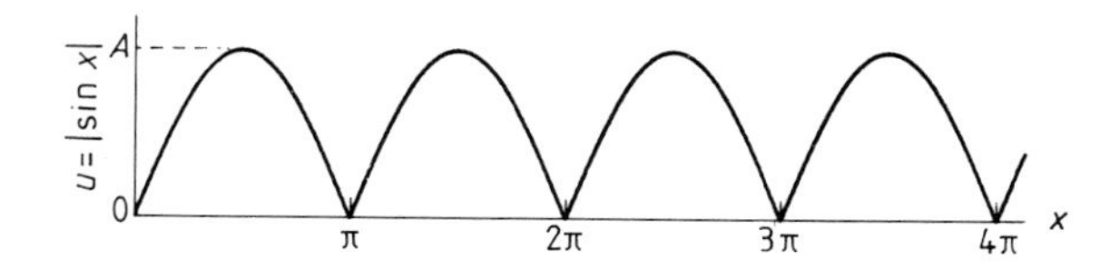

$$|\sin x|=\frac{2A}{\pi}-\frac{4A}{\pi}\left(\frac{\cos 2x}{1\times 3}+\frac{\cos 4x}{3\times 5}+\frac{\cos 6x}{5\times 7}+\cdots\right).$$

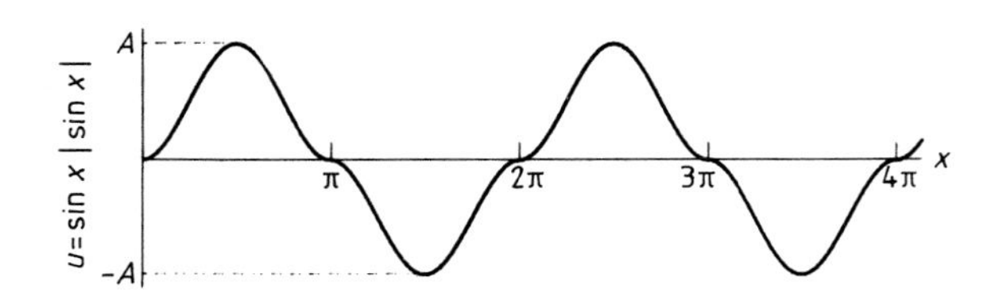

$$\sin x|\sin x| = \frac{8A}{\pi}\left(\frac{\sin x}{3} - \frac{\sin 3x}{1\times3\times5} - \frac{\sin 5x}{3\times5\times7} + \cdots\right).$$

Answers

Chapter 2

2.1 (a) 1, L. (b) 2, NL. (c) 2, L. (d) 2, L. (e) 2, NL. (f) 1, L. (g) 1, NL. (h) 2, L. (i) 2, NL.

2.2 (a) $\frac{\mathrm{d}u}{\mathrm{d}t}+\frac{b}{m}u=g$ and $\frac{\mathrm{d}n}{\mathrm{d}t}+Kn=0.$

(b) $\tau=m/b.$

(c) $u=mg/b$ and $n=0.$

2.3 (a) $\tau=1/\sqrt{(2a)}.$

(b) $f=\frac{\omega}{2\pi}=\frac{1}{2\pi\sqrt{(LC)}}.$

2.4 (a) Equations (b) and (e).

(b) Equations (c) and (i).

2.7 (a) $1+a\omega^2+a^2\omega x.$

(b) $1+a^2\omega x.$

(c) $1+a\omega^2\cos\omega x.$

2.8 (a) $\frac{\mathrm{d}^2h}{\mathrm{d}t^2}+\frac{3g}{2R}h=0.$

(b) $f=\frac{1}{2\pi}\left(\frac{3g}{2R}\right)^{1/2}.$

(c) 0.87 Hz.

2.9 (a) Curves (b), (d) and (f).

(b) Curves (e) and (c).

(c) Curve (c).

(d) Curves (b) and (f).

(e) Curve (a).

2.10 (a) $u' = -\frac{1}{l}\cos \omega x - \omega\left(1 - \frac{x}{l}\right)\sin \omega x$

$$u'' = \frac{2\omega}{l}\sin \omega x - \omega^2\left(1 - \frac{x}{l}\right)\cos \omega x.$$

(b) $\frac{\partial u}{\partial t} = \omega \cos\left(\frac{x}{l}\right)\cos \omega t \qquad \frac{\partial u}{\partial x} = -\frac{1}{l}\sin\left(\frac{x}{l}\right)\sin \omega t$

$$\frac{\partial^2 u}{\partial t^2} = -\omega^2 \cos\left(\frac{x}{l}\right)\sin \omega t \qquad \frac{\partial^2 u}{\partial x^2} = -\frac{1}{l^2}\cos\left(\frac{x}{l}\right)\sin \omega t.$$

(c) $\frac{\partial u}{\partial t} = \omega \sin\left(\frac{x}{l} - \omega t\right) \qquad \frac{\partial u}{\partial x} = -\frac{1}{l}\sin\left(\frac{x}{l} - \omega t\right)$

$$\frac{\partial^2 u}{\partial t^2} = -\omega^2 \cos\left(\frac{x}{l} - \omega t\right) \qquad \frac{\partial^2 u}{\partial x^2} = -\frac{1}{l^2}\cos\left(\frac{x}{l} - \omega t\right).$$

Chapter 3

3.1 (a) $u(0)=0$ (auxiliary condition) and equation show that $u'=0$ when $1-u^3=0$, i.e. $u'(0)=1$; u' falls as u increases.

(b) $1-u_\infty^3=0$, so $u_\infty=1$.

(c) $u^* = 1 - e^{-t/\tau}$.

(d) $\frac{1}{\tau}e^{-t/\tau} = 1-(1-e^{-t/\tau})^3 + \mathscr{R}$.

(e) $\tau = 4/7$.

3.2 (a) Equation (3.18): $u' = -au^n$, $u(0)=u_0$ and $u'(0)$ is negative ⇒ figure 3.2; $u^* = u_0\,e^{-t/\tau}$.
Equation (3.19): $u' = 1-au^n$, $u(0)=0$ and $u'(0)$ is positive ⇒ figure 3.3; $u^* = (1/a^{1/n})(1-e^{-t/\tau})$.

(b) Equation (3.18): $\tau = \dfrac{1}{a}\left(\dfrac{2}{u_0}\right)^{n-1}$.

Equation (3.19): $\tau = \dfrac{1}{a^{1/n}[2-(1/2)^{n-1}]}$.

3.3 (a) $h\theta_\infty^{5/4} = P$ (since $\theta'(\infty) = 0$).

(b) By physical or graphical argument (§3.3.2), shape is as figure 3.3; $\theta^* = \theta_\infty(1 - e^{-t/\tau})$.

(c) $\dfrac{c\theta_\infty\, e^{-t/\tau}}{\tau} + h\theta_\infty^{5/4}(1 - e^{-t/\tau})^{5/4} = P + \mathscr{R}$.

(d) Collocation gives $\dfrac{c\theta_\infty}{2\tau} + h\theta_\infty^{5/4}(\tfrac{1}{2})^{5/4} = P$; so $\tau = 0.86\,\dfrac{c\theta_\infty}{P}$.

3.4 (a) $a(\theta_\infty + 300)^4 = P$; so $\tau = \dfrac{c\theta_\infty}{2P[1 - (\frac{1}{2}\theta_\infty + 300)^4/(\theta_\infty + 300)^4]}$.

(b) $\dfrac{\tau_{\text{conv}}}{\tau_{\text{rad}}} = 1.72\left(1 - \dfrac{(\frac{1}{2}\theta_\infty + 300)^4}{(\theta_\infty + 300)^4}\right)$.

(c) $\tau_{\text{conv}}/\tau_{\text{rad}} = 1.18$.

3.5 (a) $v^* = v_\infty - (v_\infty - v_0)\, e^{-t/\tau}$.

(c) $\tau = 1.6m/bv_0^{0.7}$.

3.6 (a) Use $n^* = \dfrac{b}{2}(1 - e^{-t/\tau})$; $\tau = \dfrac{1}{Kab(1 - b/4a)}$.

(b) $\tau_c = \dfrac{5}{2}\,\dfrac{1}{Kab(1 - 2b/5a)}$.

3.7 (a) $h^* = h_\infty(1 - e^{-t/\tau})$; $h_\infty = d/g$.

(b) Eliminate d using h_∞.

(c) $v = \dfrac{R^2\tau}{8}\left(\dfrac{2g}{h_\infty} - \dfrac{1}{4\tau^2}\right)$.

Chapter 4

4.1 (a) Qualitative sketch is an exponential fall from $v = v_0$ at $x = 0$ to a plateau with $v = v_\infty$ as $x \to \infty$. TF is $v^* = v_\infty + (v_0 - v_\infty)\, e^{-x/\lambda}$.

(b) $-m(v_0\,e^{-x/\lambda})\left(\dfrac{v_0\,e^{-x/\lambda}}{-\lambda}\right)+b(v_0\,e^{-x/\lambda})^n=\mathscr{R}.$

(c) $\lambda=\dfrac{m}{b}\left(\dfrac{2}{v_0}\right)^{n-2}$; for $n=2$, $\lambda=m/b$.

4.2 (a) $x-\lambda(1-e^{-x/\lambda})$.

(b) $-\dfrac{e^{-x/\lambda}}{\lambda^2}+[x+\lambda(e^{-x/\lambda}-1)]\dfrac{e^{-x/\lambda}}{\lambda}=\mathscr{R}.$

(c) $\lambda=2.3$ (15% error from exact solution 2.0).

(d) $u^{*\prime}(0)=1/\lambda=0.43$ (30% error from exact solution 0.332).

4.3 (a) Qualitative sketch as figure 4.4(*a*) and (*b*); for $0\leqslant x\leqslant l$, $\theta^*=\theta_\infty(1-e^{-x/\lambda})$.

(b) $\lambda=1/\sqrt{a}$.

(c) Minimum length $=4\lambda=4/\sqrt{a}$.

Chapter 5

5.1 (a) $\dfrac{2\theta_c}{l^2}+a(\tfrac{8}{9}\theta_c)^n=b.$

(b) $l=\left(\dfrac{2\theta_c}{b}\right)^{1/2}\left(\dfrac{1}{1-a(\tfrac{8}{9}\theta_c)^n/b}\right)^{1/2}.$

(c) $l=\left(\dfrac{2\theta_c}{b}\right)^{1/2}+a\left(\dfrac{2\theta_c}{b}\right)^{1/2}\dfrac{(\tfrac{8}{9}\theta_c)^n}{b}.$

5.2 (a) Try to find v_∞ by putting $dv/dr=0$. This gives $p/\eta L=0$, impossible if $p\neq 0$.

(b) $v^*=v_c\left(1-\dfrac{r^2}{R^2}\right).$

(c) $v_c=\dfrac{pR^2}{4\eta L}.$

5.3 (a) $u^* = u_0\left(1 - \dfrac{4t^2}{T^2}\right)$.

(b) $T = \left(\dfrac{32}{(\frac{8}{9})^3 c u_0^2}\right)^{1/2} = \dfrac{6.75}{c u_0^{1/2}}$.

(c) $f = 0.15 c u_0^{1/2}$.

5.4 (a) Following §2.6, except that the dependent variable is y instead of u, $qy = c/y = (c/y^2)y$ so $q = c/y^2$, which is always positive.

(b) $y^* = y_0(1 - x^2/L^2)$ (c) $L = \dfrac{4y_0}{3c^{1/2}}$.

5.5 (a)

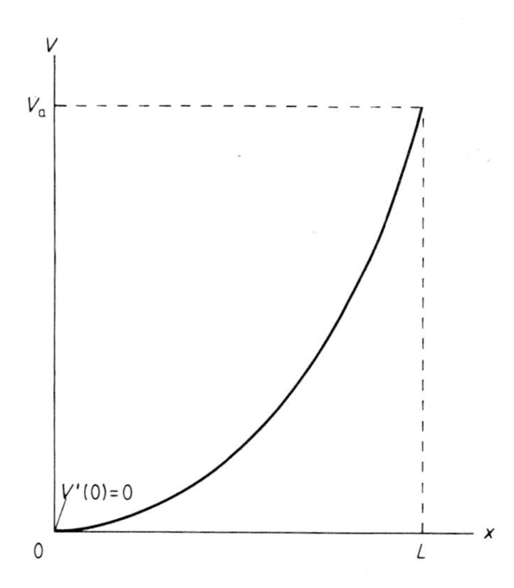

Chapter 6

6.1 (b) Integral is π/ω.

6.2 (a) $-\omega^2 u_1 \cos \omega t + c u_1^3 \cos^3 \omega t = \mathcal{R}$.

(b) $-\omega^2 u_1 \cos \omega t + c u_1^3(\frac{3}{4} \cos \omega t + \frac{1}{4} \cos 3\omega t) = \mathcal{R}$.

(c) $-\omega^2 u_1 + c u_1^3(\frac{3}{4}) = 0$; $\omega = (\frac{3}{4})^{1/2} c^{1/2} u_1$; $f = \omega/2\pi = 0.138 c^{1/2} u_1$.

6.3 (c) $T=\frac{2\pi}{\omega}=2\pi\left(\frac{l}{g}\right)^{1/2}\left(\frac{(\sqrt{3})\theta_0/2}{\sin[(\sqrt{3})\theta_0/2]}\right)^{1/2}$;

$T\to 2\pi\left(\frac{l}{g}\right)^{1/2}$ when $\theta_0\to 0$; $(\sqrt{3})\theta_0/2=180°$ when $\theta_0=208°$.

6.4 (a) $u^*=u_0+u_1\cos\omega t$. $[mg\Rightarrow mgu^0$; see §6.2]

(d) $-m\omega^2u_1\cos\omega t+\alpha(u_0+u_1\cos\omega t)$

$$+\beta[(u_0^3+\tfrac{3}{2}u_0u_1^2)+(3u_0^2u_1+\tfrac{3}{4}u_1^3)\cos\omega t]-mg=\mathscr{R}.$$

6.5 (a) $\theta''+\left(\frac{g}{l}\right)(\theta-\frac{1}{6}\theta^3)=\frac{F}{ml}(1-\frac{1}{2}\theta^2)$.

(b) $\theta^*=\theta_0+\theta_1\cos\omega t$. [§6.2]

(c) $\theta^{*3}=(\theta_0^3+\frac{3}{2}\theta_0\theta_1^2)+(3\theta_0^2\theta_1+\frac{3}{4}\theta_1^3)\cos\omega t$

$\theta^{*2}=(\theta_0^2+\frac{1}{2}\theta_1^2)+2\theta_0\theta_1\cos\omega t$.

(e) $\omega=\left(\frac{g}{l}\right)^{1/2}\left\{1-\frac{1}{16}\theta_1^2+\frac{1}{4}\left(\frac{F}{mg}\right)^2\right\}$.

6.6 (a) $\theta^{*\prime}=-\frac{kA_0}{T}\cos\omega t-A_0\omega\left(1-\frac{kt}{T}\right)\sin\omega t$

$\theta^{*\prime\prime}=\frac{2A_0\omega k}{T}\sin\omega t-A_0\omega^2\left(1-\frac{kt}{T}\right)\cos\omega t$.

6.7 (b) $\frac{2A_0\omega k}{T}\sin\omega t-A_0\omega^2\left(1-\frac{kt}{T}\right)\cos\omega t$

$$-bA_0^2\omega^2\left(\frac{8}{3\pi}\right)\sin\omega t+(g/l)A_0\left(1-\frac{kt}{T}\right)\cos\omega t=\mathscr{R}.$$

(c) Remember $\omega T=2\pi$.

Chapter 7

7.1 (a) $(Q_0-\frac{1}{2}\varepsilon Q_1\cos\phi)+(Q_1-\varepsilon Q_0\cos\phi)\cos\omega t$

$$-(\omega RC_0Q_1-\varepsilon Q_0\sin\phi)\sin\omega t$$

$$-\tfrac{1}{2}\varepsilon Q_1(\cos 2\omega t\cos\phi-\sin 2\omega t\sin\phi)=C_0V_0+\mathscr{R}.$$

(b) $Q_0-\frac{1}{2}\varepsilon Q_1\cos\phi=C_0V_0$

$Q_1=\varepsilon Q_0\cos\phi$

$\omega RC_0Q_1=\varepsilon Q_0\sin\phi$.

(e) $v = -\dfrac{\omega\,\varepsilon R C_0 V_0 \sin \omega t}{(1+\omega^2 R^2 C_0^2)^{1/2}}$.

If $\omega R C_0 \gg 1$, $v \approx -\varepsilon V_0 \sin \omega t$.
These results are correct to first order in ε.

(f) Amplitude of residual is $\varepsilon Q_1/2$. Amplitude of fundamental is Q_1. Therefore, residual ratio is $\varepsilon/2$.

7.2 (a) Residual equation is

$$\left(-\omega^2 + \frac{3u_1^3}{4}\right) \cos \omega t + \left(\frac{u_1^3}{4}\right) \cos 3\omega t = A \cos 3\omega t + \mathcal{R}.$$

7.3 (a) Residual equation is

$$A_0\left(1+\frac{kt}{T}\right)\left(-\omega^2+\omega_0^2+\frac{\varepsilon\omega_0^2}{2}\cos\phi\right)\cos\omega t$$

$$-A_0\left[\frac{2\omega k}{T}+\frac{\varepsilon\omega_0^2}{2}\left(1+\frac{kt}{T}\right)\sin\phi\right]\sin\omega t = \mathcal{R}.$$

(e) Hint: $\varepsilon^2\omega_0^4 \geqslant 4(\omega^2-\omega_0^2)^2$

$$\Rightarrow \varepsilon\omega_0^2 \geqslant 2(\omega^2-\omega_0^2) \qquad \text{or} \qquad -\varepsilon\omega_0^2 \leqslant 2(\omega^2-\omega_0^2).$$

Also $\omega^2-\omega_0^2=(\omega+\omega_0)(\omega-\omega_0)\approx 2\omega_0(\omega-\omega_0)$

for ε small.

Chapter 9

9.1 (a) $k=\omega(\mu_0/T)^{1/2}$.

(b) Qualitative sketch and collocation as figure 9.1. TF as equation (9.2).

(c) $k=\dfrac{3}{2l}\left(1-\dfrac{l^2}{19a^2}\right)^{-1/2}$.

(d) $f=\dfrac{\omega}{2\pi}=\sqrt{\left(\dfrac{T}{\mu_0}\right)\left(\dfrac{k}{2\pi}\right)}=\dfrac{3}{4\pi l}\sqrt{\left(\dfrac{T}{\mu_0}\right)}\left(1+\dfrac{l^2}{18a^2}\right)=f_0\left(1+\dfrac{l^2}{18a^2}\right)$.
(Note l/a small; f_0 is value of f when $l/a=0$.)

9.2 (a) $y^*=y_0\left(1-\dfrac{x^2}{L^2}\right)$.

(b) $k=\dfrac{9}{4}\dfrac{b^2}{L^2}\left(1-\dfrac{L}{3b}\right)^4$.

(c) 4.34.

9.3 (c) In the notation of §2.6

$$q = \frac{\omega^2}{c^2[1 - \frac{9}{8}(\mathrm{d}U/\mathrm{d}x)^2]} > 0$$

since $\frac{9}{8}(\mathrm{d}U/\mathrm{d}x)^2 < 1$ in any practical case.

9.4 (a) $\dfrac{\partial u^*}{\partial t} = U\left[-\dfrac{k}{T}\cos\omega t - \omega\left(1 - \dfrac{kt}{T}\right)\sin\omega t\right].$

9.5 (c) $U(R) = 0$; $U'(0) = 0$. [§9.6]

(e) Collocation at $r/R = \frac{1}{2}$. [End of §9.3]

Chapter 10

10.1 (a)

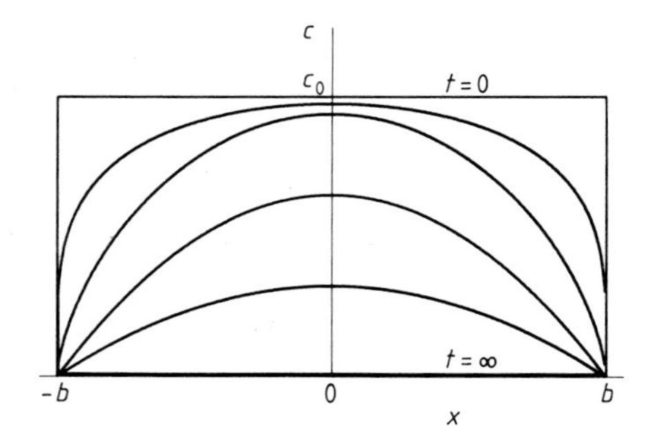

(b) $c^* = C\left(1 - \dfrac{x^2}{b^2}\right)$; $C = C(t)$.

(c) Residual equation is

$$-\frac{2C}{b^2} = \frac{1}{D}\frac{\mathrm{d}C}{\mathrm{d}t}\left(1 - \frac{x^2}{b^2}\right) + \mathscr{R}.$$

Collocation point is at $x = b/3$.

$\tau = \dfrac{4b^2}{9D}$; $t_s = 2.3\tau = \dfrac{b^2}{D}$. [See table 3.2]

(d) For $t_s = 3$ years, the old rule gives $2b = 3$ inches. Therefore

$$D = b^2/t = \tfrac{3}{4}\ \mathrm{in}^2\ \mathrm{yr}^{-1}.$$

The new rule gives

$$t_s = \frac{(2b)^2}{4D} = \frac{(2b)^2}{3}$$

or in words:

'The seasoning time in years is one-third the square of the thickness in inches.'

10.2 (a) $c^* = c_0 + [C_L(t) - c_0]\left[1 - \left(\frac{x-L}{L}\right)^2\right]$.

(b) Residual equation is

$$-\frac{2(C_L - c_0)}{L^2} = \frac{1}{D}\left[1 - \left(\frac{x-L}{L}\right)^2\right]\frac{dC_L}{dt} + \mathscr{R}.$$

Collocate at $x = 2L/3$ to get

$$\frac{dC_L}{dt} + \left(\frac{9}{4}\frac{D}{L^2}\right)C_L = \frac{9D}{4L^2}c_0.$$

(c) $\tau = \frac{4L^2}{9D}$. [See §2.3]

(d) Time to reach $0.99c$ for exponential (table 3.2) is $t_d = 5\tau = 2.2L^2/D$.

(e) Time for 2λ to reach L (inverted formula (10.18)) is

$$t_1 < \frac{L^2}{10D} \ll t_d.$$

10.3 (a) TF describes a plane temperature wave propagating into the Earth, having an amplitude that varies with depth, a wavelength $2\pi/k$ and a velocity $c = \omega/k$.

(b) Residual equation is

$$(A'' - k^2 A)\cos(\omega t - kx) + \left(2kA' + \frac{\omega A}{D}\right)\sin(\omega t - kx) = \mathscr{R}.$$

(c)
$$\frac{dA}{dx} + \left(\frac{\omega}{2kD}\right)A = 0 \qquad \frac{d^2 A}{dx^2} - k^2 A = 0.$$

(d) $A = A_0\,e^{-x/\lambda}$ $\qquad \lambda = 2kD/\omega$.

(e) $k = \left(\frac{\omega}{2D}\right)^{1/2}$

(k positive for wave into the Earth: see (a)).

(f) For exponential fall, amplitude is 10% of initial value after 2.3λ. [Table 3.2]

$$\lambda = \left(\frac{2D}{\omega}\right)^{1/2} = 0.18 \text{ m (daily)} \quad \text{or} \quad 3.5 \text{ m (annual)}$$

$$2.3\lambda = 0.41 \text{ m (daily)} \qquad \text{or} \quad 9 \text{ m (annual)}.$$

[ω is found from $\omega = 2\pi/T$ where $T = 24$ h (daily) or 365 days (annual)]

(g) Time lag $= \dfrac{\text{distance}}{\text{velocity}}$.

$$c = \frac{\omega}{k} = \frac{2\pi}{kT} = 7 \times 10^{-7} \text{ m s}^{-1}.$$

Therefore, time lag for annual variations is 1.3×10^7 s ≈ 0.4 yr.

10.4 (a)

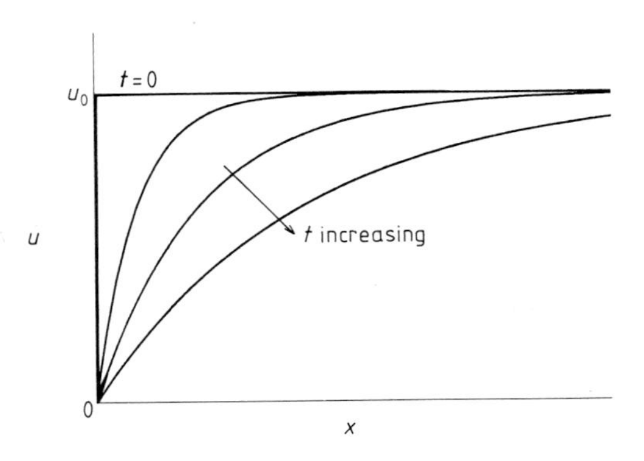

(b) $u^* = u_0(1 - e^{-x/\lambda})$ $\qquad \lambda = \lambda(t)$.

(c) Residual equation is

$$-\frac{u_0 e^{-x/\lambda}}{\lambda^2} = -\frac{x u_0 e^{-x/\lambda}}{D\lambda^2} \frac{d\lambda}{dt} + \mathscr{R}.$$

Collocate at $e^{-x/\lambda} = \frac{1}{2}$; $x = \lambda \ln 2$, to obtain ODE 1 for λ, whence

$$\lambda = \left(\frac{2Dt}{\ln 2}\right)^{1/2}.$$

(d) Gradient $= \left.\dfrac{\partial u^*}{\partial x}\right|_{x=0} = \dfrac{u_0}{\lambda} = u_0 \left(\dfrac{\ln 2}{2Dt}\right)^{1/2}$.

(e) Substitute D, the gradient, and t = age of Earth in seconds into formula in (d): $u_0 = 1.9 \times 10^4$ K. (This is greater than the boiling or decomposition point of all rocks and elements.)

10.5 (a) From equation, since $\partial^2 c/\partial x^2 \to 0$ as $x \to \infty$

$$\frac{\partial c(\infty)}{\partial t} = G; \text{ so } c(\infty) = Gt.$$

(b)

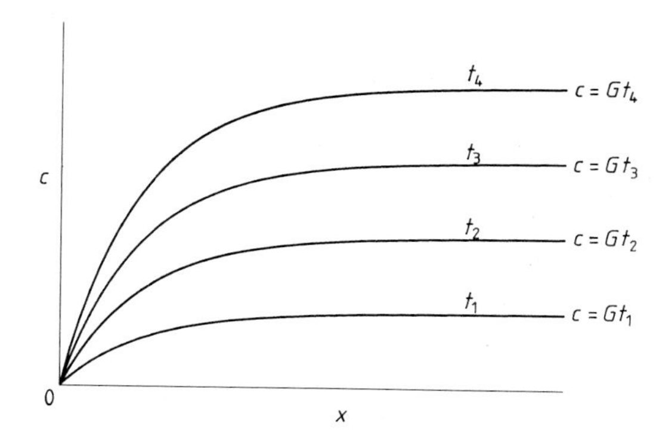

(c) $c^* = Gt(1 - e^{-x/\lambda})$ $\qquad \lambda = \lambda(t)$.

(f) $$\left.\frac{\partial c}{\partial x}\right|_{x=0} = \frac{GT^{1/2}}{D^{1/2}}.$$

$$\text{Rate of escape per unit area of surface} = D\left.\frac{\partial c}{\partial x}\right|_{x=0}.$$

So total rate of escape $= 4\pi R^2 D^{1/2} G T^{1/2}$.

Index